Hard-to-Teach SCIENCE CONCEPTS

A Framework to Support Learners, Grades 3–5

Hard-to-Teach SCIENCE CONCEPTS

A Framework to Support Learners, Grades 3–5

By Susan Koba with Carol T. Mitchell

National Science Teachers Association
Arlington, Virginia

National Science Teachers Association

Claire Reinburg, Director
Jennifer Horak, Managing Editor
Andrew Cooke, Senior Editor
Judy Cusick, Senior Editor
Wendy Rubin, Associate Editor
Amy America, Book Acquisitions Coordinator

ART AND DESIGN
Will Thomas Jr., Director, cover and interior design

PRINTING AND PRODUCTION
Catherine Lorrain, Director

SCILINKS
Tyson Brown, Director
Virginie L. Chokouanga, Customer Service and Database Coordinator

NATIONAL SCIENCE TEACHERS ASSOCIATION
Francis Q. Eberle, PhD, Executive Director
David Beacom, Publisher

Copyright © 2011 by the National Science Teachers Association.
All rights reserved. Printed in the United States of America.
14 13 12 11   4 3 2 1

LIBRARY OF CONGRESS CATALOGING-IN-PUBLICATION DATA
Koba, Susan.
Hard-to-teach science concepts : a framework to support learners, grades 3–5 / by Susan Koba with Carol T. Mitchell.
    p. cm.
Includes bibliographical references and index.
ISBN 978-1-936137-15-2
1. Science—Study and teaching (Primary) I. Mitchell, Carol T. II. Title.
  LB1585.K63 2011
  372.35'044—dc22
                    2011007668

e-ISBN 978-1-936137-45-9

NSTA is committed to publishing material that promotes the best in inquiry-based science education. However, conditions of actual use may vary, and the safety procedures and practices described in this book are intended to serve only as a guide. Additional precautionary measures may be required. NSTA and the authors do not warrant or represent that the procedures and practices in this book meet any safety code or standard of federal, state, or local regulations. NSTA and the authors disclaim any liability for personal injury or damage to property arising out of or relating to the use of this book, including any of the recommendations, instructions, or materials contained therein.

PERMISSIONS
Book purchasers may photocopy, print, or e-mail up to five copies of an NSTA book chapter for personal use only; this does not include display or promotional use. Elementary, middle, and high school teachers may reproduce forms, sample documents, and single NSTA book chapters needed for classroom or noncommercial, professional-development use only. E-book buyers may download files to multiple personal devices but are prohibited from posting the files to third-party servers or websites, or from passing files to non-buyers. For additional permission to photocopy or use material electronically from this NSTA Press book, please contact the Copyright Clearance Center (CCC) (*www.copyright.com*; 978-750-8400). Please access *www.nsta.org/permissions* for further information about NSTA's rights and permissions policies.

Featuring SciLinks®— Up-to-the-minute online content, classroom ideas, and other materials. For more information, go to www.scilinks.org/faq/moreinformation.asp.

Contents

Foreword by Linda Froschauer vii

Acknowledgments ix

About the Authors xi

Introduction xiii

Part I: The Toolbox: A Framework and Instructional Tools 1

Chapter 1. An Instructional Planning Framework to Address Conceptual Change 3
- Why Are Some Science Concepts Hard to Teach? 5
- Introducing the Instructional Planning Framework 6
- Comparing the Framework to Other Models 9
- Putting the Instructional Planning Framework Into Practice 12

Chapter 2. Implementation of the Framework Using the Topic "The Flow of Matter and Energy in Ecosystems" 17
- Overview 18
- Application of the Predictive Phase to "The Flow of Matter and Energy in Ecosystems" 20
- Application of the Responsive Phase to "The Flow of Matter and Energy in Ecosystems" 35
- Instructional Tool 2.3: Instructional Strategy Selection Tool 42
- Time for Reflection 58
- Resulting Lessons for Learning Targets #1–4 59
- Instructional Tool 2.1: Teaching the Five Essential Features of Inquiry 76
- Instructional Tool 2.2: Three Strategies That Support Metacognition 83
- Instructional Tool 2.4: Sense-Making Approaches: Linguistic Representations—Writing to Learn 90
- Instructional Tool 2.5: Sense-Making Approaches: Linguistic Representations—Reading to Learn 98
- Instructional Tool 2.6: Sense-Making Approaches: Linguistic Representations—Speaking to Learn 103
- Instructional Tool 2.7: Sense-Making Approaches: Nonlinguistic Representations—Six Kinds of Models 107
- Instructional Tool 2.8: Sense-Making Approaches: Nonlinguistic Representations—Visual Tools 114
- Instructional Tool 2.9: Sense-Making Approaches: Nonlinguistic Representations—Drawing Out Thinking 121
- Instructional Tool 2.10: Sense-Making Approaches: Nonlinguistic Representations—Kinesthetic Strategies 125

Chapter 3. The Framework and Instructional Tools at the Elementary Level 127
- Responding to the Needs of All Learners 129
- Ties to Literacy and Numeracy (Mathematics) 143
- Variations in Third, Fourth, and Fifth Grades 153
- Build Your Library 154

Part II: Toolbox Implementation—Using the Framework and Instructional Tools With Hard-to-Teach Science Topics ... 157

Chapter 4. Matter and Its Transformations: Gas Is Matter ... 159
- Overview ... 160
- Why This Topic? ... 160
- The Predictive Phase ... 161
- The Responsive Phase ... 164
- The Lessons: Teaching and Learning About "Gas Is Matter" ... 166
- Time for Reflection ... 169
- Ties to Literacy and Numeracy (Mathematics) ... 170
- Consideration Across the Grades ... 170
- Build Your Library ... 170

Chapter 5. Earth's Shape and Gravity
(by Cary I. Sneider with Susan Koba) ... 171
- Overview ... 172
- Why This Topic? ... 173
- The Predictive Phase ... 175
- The Responsive Phase ... 187
- The Lessons: Teaching and Learning About "Earth's Shape and Gravity" ... 193
- Time for Reflection ... 199
- Responding to the Needs of All Learners ... 200
- Ties to Literacy and Numeracy (Mathematics) ... 202
- Consideration Across the Grades ... 204
- Build Your Library ... 204

Chapter 6. Understanding Changes in Motion
(by Rand Harrington with Susan Koba) ... 205
- Overview ... 206
- Why This Topic? ... 207
- What Makes These Ideas Difficult? ... 208
- The Predictive Phase ... 209
- The Responsive Phase ... 214
- The Lessons: Teaching and Learning About "Understanding Changes in Motion" ... 220
- Time for Reflection ... 223
- Responding to the Needs of All Learners ... 223
- Ties to Literacy and Numeracy (Mathematics) ... 227
- Consideration Across the Grades ... 230
- Build Your Library ... 230

Appendix A: Planning Template for the Predictive Phase ... 233

Appendix B: Strategy Selection Template ... 234

Appendix C: Planning Template for the Responsive Phase ... 235

References ... 237

Index ... 249

Foreword

All elementary school teachers are aware of science content that is particularly difficult for students to comprehend. Reasons for those difficulties are varied: perhaps students have persistent preconceptions (especially misconceptions) about the content; lack previous life experiences (including those they might have missed in school) that would have provided valuable background information on the topic; have limited ability in the math skills needed for a particular subject; lack the capacity to understand abstract ideas; or require a lot of extra scaffolding by the teacher for conceptual understanding.

Addressing the needs of learners in these situations can be frustrating and difficult. Furthermore, when that content knowledge is a required student learning outcome, teachers are faced with serious instructional dilemmas. And if these difficulties are not addressed, students can develop even more misconceptions and more gaps in their learning. It is imperative, therefore, that elementary teachers explicitly instruct students in certain hard-to-understand science concepts.

In *Hard-to-Teach Science Concepts: A Framework to Support Learners, Grades 3–5*, Susan Koba and Carol Mitchell offer us a process to identify concepts that our students may not easily comprehend; the authors also provide us with the tools to create learning opportunities that can help students develop understanding. They introduce the Instructional Planning Framework and then apply it to four science topics that teachers in grades three to five can easily relate to: the flow of matter and energy in ecosystems, matter and its transformations, Earth's shape and gravity, and understanding changes in motion. These are topics that we recognize as being difficult for our students to understand (and sometimes difficult for us to understand, as well—hard as that may be to admit!).

Hard-to-Teach Science Concepts first takes teachers through the predictive phase of the framework, at which time students' preconceptions are identified and the learning targets and the learning sequence for a given unit are clarified. The responsive phase, the heart of the book, comes next. The authors show teachers how to consider the needs of their students in creating conceptual understanding by following a step-by-step process to help students gain understanding through sense making. To anchor the teacher-reader's learning and provide insight into the process, the authors walk teachers through the application of the responsive phase to the four science topics cited earlier.

Foreword

One component of this book in particular makes it an exceptional resource: its wide variety of detailed instructional tools and research-based strategies. The resources show how to recognize student needs, how to develop a concept, and how to pre- and post-assess for understanding. Furthermore, the tools and strategies are applicable to many settings. Teachers and curriculum developers will be able to apply these resources to other "hard-to-learn" teaching situations they confront. Thus, this collection of tools and strategies will be used again and again.

Too often, teachers are left to learn instructional strategies on their own and then to implement them as best they can. *Hard-to-Teach Science Concepts* gives teachers the support they need. As I read this book, I thought of a scenario in which the book's Instructional Planning Framework could impact an entire teaching team or Professional Learning Community (PLC). The format lends itself to the creation of an effective learning community that can develop and support instructional units on other science topics for an elementary teacher who may be a generalist rather than one with a considerable background in science. The book directly supports the three principles that should guide a PLC: (1) Ensure that students learn, (2) Develop a culture of collaboration, and (3) Focus on results (DuFour 2004). Beginning with a brief teacher self-preassessment and moving through suggestions for building a personal reference library and taking the next steps in professional development, this book provides all of the tools necessary for an effective PLC.

Susan Koba and Carol Mitchell address themselves to the reader who is thinking about teaching—or is required to teach—a hard-to-teach concept: "Start this process by being very clear about why you are teaching what you are teaching." The authors warn that "you cannot assume that just because the topic is covered in your textbook or science kit that this coverage truly addresses the big ideas." In their book, on the other hand, simple, clear advice in the many charts, figures, templates, and diagrams is coupled with explanatory text that shows the teacher just how to apply the framework. We have here a perfect learning tool for change. I know teachers will join me in thanking Susan Koba and Carol Mitchell for this valuable resource.

Linda Froschauer
Editor, Science and Children
NSTA President 2006–2007

DuFour, R. 2004. Schools as learning communities. *Educational Leadership* 61 (8): 6–11.

Acknowledgments

Susan Koba, primary author, would like to thank each of the following individuals for their contributions and strong support during the research for and writing of Hard-to-Teach Science Concepts: A Framework to Support Learners, Grades 3–5.

My sincerest thanks go to Dr. Carol T. Taylor Mitchell—my mentor, colleague, and longtime friend—for her elementary instruction perspectives, contributions to the book, and guidance during the writing process.

Special thanks go to Judy Cusick, the NSTA Press editor for my first and for what is now my second book. I could not ask for a more supportive and flexible editor.

I greatly appreciate the students in Dr. Mitchell's graduate courses who responded to a survey designed to identify hard-to-teach science concepts. I also give special thanks to two of her preservice teachers, Jacquelyn Sawaged and Lynn Schreck, who took time from their busy lives as students to review chapters and provide feedback.

Most important, I want to express my deepest appreciation for my husband, Mike, and his continued support and patience during the writing process.

About the Authors

Susan Koba, a science educator for almost 40 years, now serves as a science education consultant. The majority of her consulting is with the National Science Teachers Association (NSTA), where she works on projects related to the NSTA Learning Center. She is a past director of coordination and supervision on the NSTA Board of Directors, past member of numerous NSTA committees and tasks forces, and past president of her state NSTA chapter. She is currently president of the National Science Education Leadership Association.

Susan retired from the Omaha Public Schools (OPS) after 30 years. She taught both middle and high school science, eventually becoming a curriculum specialist and district mentor. She closed her career with OPS as the project director and professional development coordinator for the National Science Foundation–funded Banneker 2000: CEMS [Community of Excellence in Mathematics and Science] Urban Systemic Program initiative at OPS and the University of Nebraska-Omaha.

Susan's contributions to science education have been recognized through many awards. She has been named Outstanding Biology Teacher for Nebraska, Tandy Technology Scholar, Genentech Access Excellence Fellow, and Christa McAuliffe Fellow and has received the Presidential Award for Excellence in Science Teaching. She received her BS in biology and secondary education from Doane College, her MA in biology from the University of Nebraska-Omaha, and her PhD in science education from the University of Nebraska-Lincoln.

Carol T. Mitchell, a public educator for the past 40 years, is a professor of teacher education at the University of Nebraska-Omaha (UNO), where she teaches Science Methods, Trends and Issues in Science Education, and Contemporary Issues in Science Education. Carol's reach has extended well beyond Nebraska: She is an alumnus of the 2002 and 2009 Oxford Roundtables, she conducted an educator's conference for K–12 math and science teachers in Swaziland and Lesotho, and she conducted the first Summer Science Institute for teachers and students in Maseru, Lesotho. Prior to her work at UNO, she taught science at the middle and high school levels and was the science supervisor for the Omaha Public Schools (OPS), the largest urban school district in Nebraska.

Carol was the coprincipal investigator of the National Science Foundation–funded Banneker 2000: CEMS [Community of Excellence in Mathematics and Science] Urban

About the Authors

Systemic Program initiative at OPS and UNO. Mitchell was also the co-principal investigator of the Omaha Public Schools Comprehensive Partnership in Mathematics and Science, which sought to increase achievement in mathematics and science among African American students.

She consults for major book companies and sits on the editorial board of the *Journal of African American Studies.* Carol holds a BS in secondary education from North Texas State University, an MS in chemistry from Southern University, an administrative certification from UNO, and a PhD in curriculum and instruction from the University of Nebraska-Lincoln. Her various professional recognitions include the UNO Alumni Outstanding Teacher Award, the UNO Excellence in Teaching Award, the Western Heritage Museum's Outstanding African American Achievement Award, and the Barrier Breaker Award from the University of North Texas, which she received at the university's celebration of the 50th anniversary of desegregation.

* * * * * * *

Chapter Contributors

Chapter 5: "Earth's Shape and Gravity"
Cary I. Sneider
Associate Research Professor
Portland State University
Portland, Oregon

Chapter 6: "Understanding Changes in Motion"
Rand Harrington
preK–12 Science Department Chair and Science Curriculum Coordinator
The Blake School
Visiting Assistant Professor at the University of Minnesota
Minneapolis, Minnesota

Introduction

> "Except for children (who don't know enough not to ask the important questions), few of us spend much time wondering about why nature is the way it is; where the cosmos came from, or whether it was always here; if time will one day flow backward and effects precede causes; or whether there are ultimate limits to what humans can know.... In our society it is still customary for parents and teachers to answer most of these questions with a shrug."
>
> —*Carl Sagan, in Hawking 1988, p. ix*

Science is a subject that warrants study and understanding by all learners as early in their schooling as possible. Research indicates over and over that children have a natural curiosity that is critical in science (Bruner and Haste 1987; Carey 1985; Dewey 1902; Eyon and Linn 1988; Martin 2006). Carl Sagan captures this idea in the above quote. So how can teachers tap into this natural resource more effectively?

In this book, *Hard-to-Teach Science Concepts: A Framework to Support Learners, Grades 3–5*, the authors use their cumulative experiences of more than 70 years of classroom teaching to share ways that teachers can make the most of children's curiosity. Unfortunately, as we know, science education has been minimal or totally lacking in U.S. elementary schools. One reason for this is the limited preparation in science received by many elementary school teachers in colleges of education, whose job it is to prepare teachers to teach many disciplines in addition to science. A second reason is that the time allocated for science during the elementary school day varies and is usually negligible (Michaels, Shouse, and Schweingruber 2008).

The good news is that over the past 10 years more attention has been given to these problems in the elementary grades. Increasingly, inservice professional development to effectively teach science is being provided to elementary school teachers. In addition, postsecondary institutions have been enhancing their programs of study for elementary majors relative to science content courses.

Hard-to-Teach Biology Concepts: A Framework to Deepen Student Understanding (Koba with Tweed 2009) addressed topics that were considered difficult to teach in high school biology courses. Stressing the importance of understanding how students learn, that

Introduction

book introduced the Instructional Planning Framework and included instructions and resources for implementing the framework.

This book is for elementary teachers, who, as we have noted, must teach all subjects, science among them. It is so important for elementary classroom teachers to increase their focus on science and to understand more fully how elementary students learn. The intent of this book is to support you in both those efforts. The same framework found in *Hard-to-Teach Biology Concepts* is used in this book. Here, the chapters that describe implementation of the framework, as well as Chapters 4–6 (plus parts of 2 and 3) that are related to content typically taught in elementary schools, were specifically developed to support elementary teachers in grades 3 through 5 and those who work with them.

This book is not a container of answers, or "fixes," to the many questions that elementary teachers have about teaching science. But it does provide the Instructional Planning Framework (Fig. 1.2, p. 8); explanations about how to use the framework; numerous figures and tables and 10 Instructional Tools (pp. 42–44 and pp. 76–126), which contain many references; and examples of application of the framework. In addition, Chapters 3–6 have extensive lists of print and online resources in the sections called Build Your Library.

Both novice and veteran teachers can use the framework and tools to develop students' conceptual understanding of four hard-to-teach and hard-to-learn science topics: a life science topic, *the flow of matter and energy in ecosystems;* two physical science topics, *matter and its transformations* and *understanding changes in motion;* and an Earth and space science topic, *Earth's shape and gravity.* You do not need to completely change what you already do to implement these ideas. However, by using the examples and the resources in the content chapters (Chapters 4–6) and parts of Chapters 2 and 3 as guides, you should be more comfortable and confident as you approach the teaching of science in what will perhaps be a new way for you. Tackle one concept or topic at a time, reflect more on your teaching, and make adjustments to your planning and teaching. This process should result in improved conceptual understanding for all your students.

Science Education Reform and Conceptual Understanding

The National Science Education Standards (NRC 1996) provide guidelines for the content to be taught and some ideas about successful instruction. Research cited in *How Students Learn: Science in the Classroom* (Donovan and Bransford 2005) extends the thinking found in the standards and reinforces the ideas that teaching science is not "telling" science and that instructing students in the scientific method and having them follow the steps is not a way to develop understandings and knowledge of how to do science. The authors of *How Students Learn* also emphasize that science educators have not done enough to support metacognitive instruction to help students learn how to learn. More recently, *Ready, Set, Science! Putting Research to Work in K–8*

Introduction

Science Classrooms (Michaels, Shouse, and Schweingruber 2008) and *Taking Science to School: Learning and Teaching in Grades K–8* (Duschl, Schweingruber, and Shouse 2009) reported on research that shows that K–8 learners come to school with far more capacity to learn science than was previously thought. With this research in hand, this book provides a framework so that elementary teachers can begin to focus not only on *what* is taught in science classes but also, and as important, on *how* it is taught.

Difficult Topics—*Why* Are They Hard to Teach and Learn?

If you reflect back on science lessons you have taught, we are sure that you can develop a list of topics that you found difficult to teach. Among the hard-to-teach topics reported in science education literature and by teachers themselves, many fall into the physical science category. (Two of those topics in physical science are addressed in this book: Chapter 4, matter and its transformations, and Chapter 6, understanding changes in motion.)

Chapter 1 discusses the possible reasons that certain topics are difficult to teach. Among these reasons is the fact that the more abstract a science topic is, the harder it is to learn for many people, including teachers! As we noted, telling science to students is not teaching science. Students learn by "doing" science, and abstract topics need to be made concrete. Students are better able to face their misconceptions and preconceptions when they are engaged in instructional activities that put science into a context they can understand.

Elementary teachers may also find physical science topics hard to teach because of the limited science content courses they took in elementary science programs of study (Michaels, Shouse, and Schweingruber 2008), as well as limited mathematics preparation. Elementary teachers may have weak areas of concentration in their backgrounds, but they have little time when they are teaching to focus on these weak areas. Science, more than other subjects, is likely to be a weak area.

We chose the hard-to-teach topics for this book in two ways: by examining the research on elementary science teaching and by determining topics that we believe are foundational to future learning. As noted, we selected four topics: one in life science, two in the physical sciences, and one in Earth and space science.

Organization of the Book

Part I. The Toolbox: A Framework and Instructional Tools

This section shares the authors' research-based framework to address conceptual change (the Instructional Planning Framework), as well as a process and Instructional Tools to put the framework into practice.

Chapter 1 explores further why some topics are hard to teach and learn, presents the Instructional Planning Framework and the research behind it, compares the

Introduction

framework to other models, and helps the reader understand how to transition from the theoretical framework to practice.

In Chapter 2, implementation of the framework and tools is modeled, based on a life science topic, the flow of matter and energy in ecosystems. Instructional Tools (pp. 42–44 and 76–126) outline various instructional approaches and specific instructional strategies that promote student understanding. By *approach,* we mean a broad method used in instruction (e.g., writing as a linguistic approach and drawing as a nonlinguistic approach). The chapter addresses metacognitive and standards-based approaches as well as various linguistic and nonlinguistic approaches. *Strategies* are specific ways to achieve a specific goal (e.g., the use of concept cartoons is one specific strategy discussed in the section on drawing as a nonlinguistic approach).

Chapter 3 provides a general overview of differentiation in the classroom and discusses numerous introductory steps for differentiating science instruction in grades 3, 4, and 5. The chapter continues to focus on the flow of matter and energy in ecosystems as these ideas are explored. It also makes connections among science, literacy, and numeracy (mathematics) and provides suggestions for their integration into the science classroom.

Part II. Toolbox Implementation—Using the Framework and Strategies With Three Hard-to-Teach Science Topics

This section is organized to model the use of the Instructional Planning Framework in three additional hard-to-teach elementary science topics: matter and its transformations, Earth's shape and gravity, and understanding changes in motion. The contributing authors of these chapters studied the framework and tools and interpreted them in their own ways. Each chapter does follow basically the same format, however, and includes three or more sample lessons on the topic.

The Five Natural Learning Systems

The five natural learning systems—emotional, social, physical, reflective, and cognitive (Given 2002; Gregory and Hammerman 2008)—are referred to frequently in this book. Chapter 3 discusses the emotional and social learning systems required to establish a safe learning environment and to engage each student in the learning process. The physical system is addressed through the active learning typical of inquiry and several strategies that are provided in Instructional Tool 2.10 (p. 125). The reflective system (metacognition) is addressed through the metacognitive tools; in fact, the process outlined in Chapter 2 calls for teachers to identify a metacognitive goal for *each* instructional unit.

The book focuses most deeply on the cognitive system, as developed in the conceptual change model (Chapter 1) and the vast array of research-based strategies. The strategies help you remain aware of your students' understandings about the targeted content before and during instruction, elicit and confront those conceptions during a lesson, and provide sense-making experiences to move student conceptions closer to the scientific explanations you expect of them. Using the framework, process, and tools will help you effectively address all five learning systems.

PART I
The Toolbox: A Framework and Instructional Tools

Chapter 1

An Instructional Planning Framework to Address Conceptual Change

CHAPTER 1: An Instructional Planning Framework to Address Conceptual Change

> "It is important for science educators to balance a deep appreciation of what is genuinely conceptually difficult, 'non obvious' and novel about many central principles of modern science, with an equally deep appreciation of the many intellectual resources that children bring to the science learning task."
>
> —*Duschl, Schweingruber, and Shouse 2007, p. 82*

Consider the various science topics that you teach during the course of the school year. Are some harder for your students to learn and for you to teach? Your answer is probably "yes" if you are like many science teachers. Yet it is your responsibility to effectively teach these topics and provide students with a solid foundation for additional science learning—both while they are still in school and when they go out into the world.

Your students' understandings about science are important partly because some of those students will become scientists. But the many who do not pursue science as a vocation will still use their knowledge of science to make informed decisions as adults and in their careers. More and more, our nation depends on technical and scientific abilities, and science is central to our culture. In addition, science provides a great context for your students to develop language, logic, and problem-solving skills (Duschl, Schweingruber, and Shouse 2007). Students' science experiences in your classroom set a foundation for their future content understandings and their comfort with science as a way of thinking. Thus, the choices you make in planning for science instruction are extremely important.

To be an effective teacher, you must focus on standards and on identifying what students should know and be able to do in core science areas. This is the case with existing standards and will continue to be important with the emerging Next Generation Science Education Standards (NRC 2010). These reforms call on you to place more emphasis on learning important concepts than on rote learning. This is not always an easy task, especially for the hard-to-teach topics and for concepts that even you might struggle to understand. The Instructional Planning Framework, on which this book is based (the framework was developed by author Susan Koba with Anne Tweed; see *Hard-to-Teach Biology Concepts: A Framework to Deepen Student Understanding* [2009]), incorporates research findings and implications for teaching, including teaching the four hard-to-teach concepts that are the focus of this book: the flow of matter and energy in ecosystems, matter and its transformations, Earth's shape and gravity, and understanding changes in motion.

CHAPTER 1: An Instructional Planning Framework to Address Conceptual Change

Why Are Some Science Concepts Hard to Teach?

Learning science is hard for many students, a reality that can make your job quite difficult. Why are certain science concepts particularly challenging for some students (or even for all students)? See Table 1.1 for possible reasons.

Table 1.1

Reasons Why Certain Science Concepts Are Hard to Teach

Abstract Concepts in Science	Students can't visualize or directly study some concepts. In their attempts to develop more concrete models, misconceptions occur.
Complex Systems in Science	The studied system is very complex and has many interacting pieces and parts. These often include multiple levels of organization (e.g., events in cells influence the functions of tissues, organs, organisms, and even the environment).
Students' Limited Previous Experiences	Foundational knowledge may be missing due to lack of prerequisite learning experiences.
Students' Limited Symbolic Tools	Students may have had limited access to symbolic tools (e.g., mathematical tools).
Persistence of Misconceptions Among Students	Some concepts are simply counterintuitive and difficult to change.

Some scientific phenomena are difficult to study through direct observations, so your students might say that the ideas are just too hard. Indeed, some ideas are very conceptual and students cannot visualize what is happening. For example, consider your students learning about the properties of air, a common expectation in third–fifth grade classrooms. They are expected to understand that air is made of matter even though they cannot see the pieces of matter with the naked eye.

Often the *system* being studied is very complex, and your students might struggle to understand the ideas without understanding the pieces and parts of the system and how they interact. Their preconceptions are likely to include incomplete (and perhaps incorrect) foundational knowledge that causes them to struggle to understand a complex science concept. Take a look at Figure 1.1, page 6, which shows a food chain typical of those used in elementary classrooms. Simple word memorization of the parts of this system does not build an understanding of its complexity. Students must be able to integrate ideas about feeding and energy, based on an understanding of what food provides for living systems. If they do not understand the scientific meaning of *food*, they will not fully understand what is represented by the food chain.

CHAPTER 1: An Instructional Planning Framework to Address Conceptual Change

Figure 1.1

Sample Representation of a Food Chain

Grass ↓ Grasshopper ↓ Frog ↓ Snake ↓ Hawk

Also consider that students learn about science in many ways through their own experiences in the natural world (e.g., their yards, the schoolyard, parks, or national forests), watching television, surfing the web, and interacting with family members. Past opportunities for rich learning experiences will vary dramatically among your students, and some of them may have limited foundational knowledge. Remember that these limitations are experiential rather than developmental; you are the person who can provide experiences to address these gaps.

In some cases, mathematical tools and models are essential to inquiry and are required for students to be able to pose questions, gather data, construct explanations, and communicate results (NRC 1996). Again, some of your students have only limited abilities to use such tools and models (NSTA 2004). Furthermore, because the typical elementary curriculum emphasizes literacy in reading and writing, often ignoring "visual literacy" skills, some student struggle in the areas of visual literacy, perhaps seeing graphs as pictures in themselves rather than representations of interactions (Vasquez, Comer, and Troutman 2010).

Although there are concepts with which young learners struggle, recent research (Duschl, Schweingruber, and Shouse 2007) shows that their thinking is quite sophisticated and that they have significant understanding of the natural world on which we can build. These findings counter earlier suggestions that young children are strictly concrete thinkers or that they are unable to learn about scientific thinking (Duschl, Schweingruber, and Shouse 2007). So do not let the difficulty of some concepts prevent you from helping students learn. The Instructional Planning Framework, the 10 Instructional Tools in Chapter 2, and the many tables and figures in this book are designed to help you find ways to work with your students' various experiences and abilities so that they can learn even the most hard-to-learn and hard-to-teach topics.

Introducing the Instructional Planning Framework

Your students develop their own ideas and explanations for concepts, building mental models (conceptual frameworks) that reflect their understandings. They may have solid foundational ideas, in which case your task is to provide experiences that your students can make sense of and incorporate into their existing ideas (conceptions). On the other hand, their existing thinking may contain faulty reasoning that they developed earlier

CHAPTER 1: An Instructional Planning Framework to Address Conceptual Change

in their lives, even in prior science classes. In this case, you must help them reconstruct their ideas and revise their mental models to better align with scientific explanations.

To help you in either of these efforts, this chapter introduces our research-based[1] Instructional Planning Framework (Figure 1.2, p. 8) (the details of how to put the framework into practice are found in Chapter 2). You can see that the Instructional Planning Framework consists of two components: the *predictive phase* and the *responsive phase*.

What Is the Predictive Phase?

Refer to the four white boxes across the top of Figure 1.2, page 8. These aspects of the framework make up the *predictive phase*—during it, you *predict* a logical instructional sequence to develop essential understandings, the conceptual target of the instructional unit. This phase is carried out *before* the school year or unit begins. (Sometimes this phase is carried out by the local school district rather than by individual teachers.) You should first clarify the essential understandings you want your students to develop. These understandings are the big ideas, central to the targeted topic and to science. If students understand these ideas, they will have a firmer foundation on which to continue learning science.

Often, however, there is a large gap between your students' ideas and the essential understandings you want them to acquire. You must identify the knowledge and skills for students to learn and identify the steps (the learning targets) in the learning sequence needed to build student understanding. These learning targets break down the essential understandings into manageable chunks for students to learn—they lessen the gap between what students currently know and the essential understandings you want them to gain. Finally, you establish the criteria that you will use to demonstrate students' conceptual understandings so that students are clear about what successful performance looks like. To measure progress to the desired outcome, you use formative assessments as checkpoints for your students and provide feedback so that they improve their understandings. And eventually you will use the criteria to develop your summative assessments.

Naturally, you need to know and understand the content yourself so that you are certain you know what your students should learn. "By taking the time to study a topic before planning a unit, teachers build a deeper understanding of the content, connections, and effective ways to help students achieve understanding of the most important concepts and procedures in that topic" (Keeley and Rose 2006, p. 5). This book includes many sources of content knowledge, including the Build Your Library sections at the end of Chapters 3, 4, 5, and 6.

What Is the Responsive Phase?

Whereas the predictive phase is research-based and developed outside of the classroom, the *responsive phase* implements a research-based plan and *responds* to your students' ideas. Keeping student conceptions in the forefront is core to this phase and is addressed through a four-step iterative process seen in Figure 1.2, on next page.

CHAPTER 1: An Instructional Planning Framework to Address Conceptual Change

Figure 1.2
Instructional Planning Framework

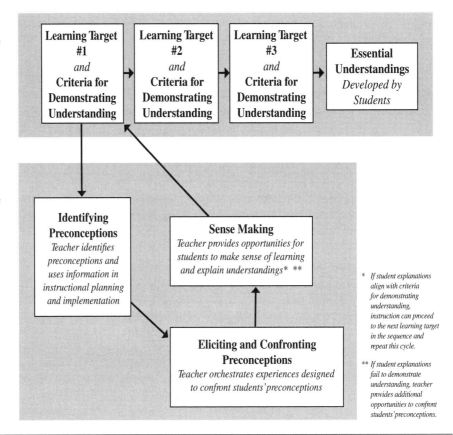

Predictive Phase
The teacher determines the lesson's essential understandings, the sequence of learning targets that lead toward those understandings, and the criteria by which understanding is determined.

Responsive Phase
Building on the foundation of the predictive phase, the teacher plans for and implements instruction during the responsive phase, one learning target at a time.

Note: This figure will be referred to throughout the book. You might want to flag it to make it easy to find.

1. Identifying Preconceptions

The importance of long-term planning, based on your students' needs, cannot be overstated. You begin this process by studying the research about common student misconceptions. This step is completed prior to instruction. Next you identify your own students' preconceptions. Based on this information, you then outline instruction focused on the conceptual target and select strategies that promote the conceptual change process.

2. Eliciting and Confronting Preconceptions

Once you have your instructional unit planned, you are ready to teach. You begin by eliciting your students' preconceptions (a time when they make their preconceptions public to their peers and you; this also might be the first time that the students themselves become aware of their preconceptions). You discuss those preconceptions with them and provide experiences that confront their conceptions (especially their misconceptions), pose alternative explanations, and encourage them to question their existing ideas. An inquiry approach is a helpful one at this stage; you will find inquiry-based instruction in each of the chapters in Part II of this book.[2] This stage may require you to make some modifications to your initial plan, but it should serve you reasonably well because now it uses both the misconceptions of many children and the preconceptions of your own students.[3]

3. Sense Making

The experiences you design to confront students' preconceptions will not, by themselves, alter student understandings to align more with scientific explanations. You need to provide opportunities for your students to make sense of the ideas they learn because it is unlikely that they will draw the appropriate conclusions on their own. You must select strategies that engage students through questioning, discussing, and other methods so that they can make connections between their ideas and what they are meant to learn. Later, you will ask students to think about their initial ideas, relate them to their current learning, and then determine how their thinking has changed.

4. Connecting Learning to the Criteria for Demonstrating Understanding (represented in Figure 1.2 by the arrow from the Sense Making box to Criteria for Demonstrating Understanding in the predictive phase)

Student learning must match the criteria for developing student understanding. If it does not, you will need to revisit the steps in the responsive phase and select additional strategies to help students understand the concepts. This is not a linear learning process; rather, it is iterative in nature to ensure that each of your students learns core science ideas.

Comparing the Framework to Other Models

The Instructional Planning Framework is partly built on Strike and Posner's (1985) Conceptual Change Model (CCM). That model says that students' conceptions can change if students are aware of their personal conceptual understandings, are introduced to new evidence that leaves them dissatisfied with their current views, and are willing to accept these new conceptions (scientific viewpoints) as more plausible than their previous conceptions. Successful use of the framework depends on understanding how to make your students' thinking visible and then either adding to their thinking or helping them reconstruct their ideas. With the framework, you'll be able to assist students to generate links between their prior knowledge and new ideas. Rather than supplying your students with information, your job becomes to facilitate students' construction of webs of connected ideas.

CHAPTER 1: An Instructional Planning Framework to Address Conceptual Change

Many of you are familiar with the well-known BSCS 5E Instructional Model (Bybee 1997) and may even use a curriculum built upon it. If that is the case, the framework and strategies found in this book serve as a nice supplement to your curriculum. If you use a more traditional curriculum, then this book will help you modify the curriculum to promote inquiry and your students' conceptual understanding. The Instructional Planning Framework incorporates and builds on both the 5E model and the Conceptual Change Model. Table 1.2 compares the framework to those models.

Table 1.2

A Comparison of the 5E Instructional Model, the Conceptual Change Model, and the Instructional Planning Framework

5E Instructional Model[a]	Conceptual Change Model[b]	Instructional Planning Framework
		Predictive Phase: Clarifies the conceptual target (essential understandings for the lesson), identifies and sequences learning targets, and provides criteria for demonstrating understanding
		Responsive Phase—Identify Preconceptions: Provides input for the teacher about students' *pre-lesson* conceptions and aids in lesson development
Engage Phase: Engages students in the topic and motivates them; identifies current conceptions; connects learning to past lessons	Commit to an Outcome: Students are expected to make predictions about an activity, which makes them aware of their preconceptions.	Responsive Phase—Elicit and Confront Preconceptions: Requires students to make evident their thinking about a concept *during instruction* and compares their ideas with those of others through various learning experiences that include classroom inquiries
Explore Phase: Provides students with a common context of shared experiences connected to the content; lets them test their own ideas and compare them to those of their peers	Expose Beliefs: Students share ideas in small groups and with the whole class.	
	Confront Beliefs: Students are given opportunities to test their ideas and discuss them in small and large groups.	

CHAPTER 1: An Instructional Planning Framework to Address Conceptual Change

Table 1.2 *(continued)*

5E Instructional Model[a]	Conceptual Change Model[b]	Instructional Planning Framework	
Explain Phase: Allows students to demonstrate their conceptual understandings	Accommodate the Concept: Students try to resolve the differences between their own ideas and their observations or what they learned, further clarifying their understanding.	Responsive Phase—Sense Making	*Perceiving, interpreting, and organizing information* Asks students to use information from learning experiences to clarify their understandings and resolve discrepancies among explanations
Elaborate Phase: Challenges and extends students' conceptual understandings by providing new experiences	Extend the Concept: Students are given opportunities to connect their understandings to new experiences.		*Connecting information* Provides additional experiences for students and asks them to connect thinking to these experiences
	Go Beyond: Students pursue related questions and problems.		*Retrieving, extending, and applying information; using knowledge in relevant ways*
Evaluate Phase: Provides opportunities for teachers to evaluate student understanding and for students to self-assess		Responsive Phase—Demonstrating Understanding: Engages students in peer- and self-assessments. Engages teachers in formative assessments that inform further instruction to guide development of student understanding.	
		Responsive Phase (revisited): If student explanations fail to demonstrate understanding, the teacher provides additional opportunities for students to confront their preconceptions.	

[a] Bybee, R. W. 1997. *Achieving scientific literacy: From purposes to practices.* Portsmouth, NH: Heinemann.
[b] Posner et al. 1982. Accommodation of a scientific conception: Toward a theory of conceptual change. *Science Education* 66 (2): 211–227.

Notice that all three models build on students' current conceptions and provide experiences that let students express their thinking and that also challenge their thinking. Each model also clarifies student understandings and then builds on and extends their thinking with new experiences. The Instructional Planning Framework differs from the other two models in its predictive phase, during which teachers establish clear learning goals, learning sequences, and criteria to demonstrate understanding—all prior to instruction. The components of the responsive phase of the Instructional Planning Framework occur in the classroom just as do the components of the 5E Instructional Cycle and the Conceptual Change Model.

Another difference, and unique to this book, is the rich set of Instructional Tools on pages 42–44 and 76–126 that are used in the process of putting the framework into practice. The Instructional Planning Framework builds strongly on conceptual change research, but also recognizes the importance of social interaction using language (discourse) and other multiple resources (e.g., writing, visual images, and physical actions) to mediate learning. Tools to facilitate social interaction are central to the sense-making approaches (e.g., see Instructional Tool 2.6). It is also important that students participate in experiences that resemble those of scientists; so tools appropriate to this task are also included (e.g., Instructional Tool 2.1). The strategies discussed in Chapter 2 both support the Instructional Planning Framework and reflect aspects of each of these three perspectives (i.e., conceptual change, social interactions, and scientist-like experiences). Thus, you can select tools that best fit both the context of your instruction and your personal teaching philosophy.

Putting the Instructional Planning Framework Into Practice

In Chapter 2, a detailed outline of how to put the framework into practice and the 10 Instructional Tools give you effective methods to support student learning. However, the framework itself is an iterative process and many of the steps can either lead you to the next step or return you to a previous step. Whether you move on or revisit previous steps depends on the depth of your students' understandings. These steps cannot be prescribed for you, but the book does outline general suggestions for key steps in planning. Use the framework and Instructional Tools in ways appropriate to determine next steps in the learning for *your* students because the context in which you teach is unlikely to be the same as that of other teachers.

The book is designed to help you not only learn about ways in which to promote your students' learning but also reflect on your own learning. Start this process by asking yourself questions such as these:

- How have I taught the content in the past?
- What do I currently know about student preconceptions and how to determine them?
- What do I know about planning and delivering instruction that engages students intellectually with the content in ways that both confront their existing ideas and help them make sense of their learning?
- How can I apply what I learn from this book to my teaching?

Refer to Table 1.3 and take a moment to respond to each statement. Then use your responses to set goals that will frame your study of this book. By focusing on the areas where you think you need to learn, your use of the book will be greatly enhanced.

CHAPTER 1: An Instructional Planning Framework to Address Conceptual Change

Table 1.3

Personal Pre-Assessment: Goals and Reflections to Guide Your Reading of This Book

Statements of Understanding Related to the Instructional Planning Framework Components	Degree of Agreement With Each Statement (Mark one of these columns with an "X" after each statement to show your level of agreement.)				
	Strongly Disagree	Disagree	Uncertain	Agree	Strongly Agree
I am comfortable identifying and using various metacognitive strategies to promote student learning.					
I am comfortable identifying and using strategies that use inquiry and the nature of science to promote students' content understandings.					
I am familiar with common misconceptions and/or how to find research on misconceptions about a topic.					
I know a variety of ways in which to determine my students' preconceptions and use those ideas to plan my lessons.					
I know a variety of effective ways to elicit and confront student preconceptions *during* my lessons.					
I am familiar with the various ways to help students make sense of the experiences used to confront misconceptions.					
I am comfortable (a) using a variety of formative assessment strategies to determine the current level of student understanding in my class and (b) using that information to modify instruction.					
I can readily provide additional experiences to readdress a concept when students do not understand it after initial instruction.					

Based on your responses above, identify two learning goals you will address as you read this book.

Learning Goal 1:

Learning Goal 2:

Hard-to-Teach Science Concepts

CHAPTER 1: An Instructional Planning Framework to Address Conceptual Change

The topics selected for inclusion in this book—the flow of matter and energy in ecosystems, matter and its transformations, Earth's shape and gravity, and understanding changes in motion—are those for which there is clear research about students' ideas and learning challenges. Once you become sufficiently adept at using the Instructional Planning Framework, you will be able to apply it to other concepts that you teach. For now, the greatest expected outcome of the use of this book is improved student understanding of four complicated science concepts.

Endnotes

[1] Both the predictive and responsive phases are grounded in research. Research addresses the importance of the following three aspects of the predictive phase: (1) identifying essential learning(s) or learning goals in a topic; (2) focusing in depth on the sequence of ideas in the related learning targets; and (3) determining criteria for demonstrating understanding for each of the learning goals (Bransford, Brown, and Cocking 1999; Heritage 2008; Masilla and Gardner 2008; Michaels, Shouse, and Schweingruber 2008; Vitale and Romance 2006). Providing students with clear learning goals is a critical first step for teachers and students and can lead to increased student achievement (Marzano, Pickering, and Pollock 2001). If you are not clear about the concepts and subconcepts that lead to science literacy, then your students will be unsure about what they are supposed to learn and will cling to facts and vocabulary without developing the fundamental understanding of the concepts. It is important to share with your students not only the standards for which they are accountable but also the subgoals embedded in those standards, arranged in a sequence that will help them progress to the ultimate learning goal.

Unfortunately, most state science standards do not provide clear pathways to understanding (Heritage 2008). Some standards don't even make it clear what a student should learn. Learning sequences give students a pathway to the ultimate learning goals or big ideas we want them to learn. Students need short-term goals so that the gaps between their current preconceptions and the desired learning goals are not too great. In the case of gaps that are too big, students tend to give up. With gaps that are too small, on the other hand, students will say that they already "get it" and are bored.

Identifying the short-term goals should also be accompanied by determining the criteria for conceptual understanding. When students are aware of your criteria, they know the evidence of learning they need to show to move forward to the next short-term goal.

Teachers are experts and students are novices in terms of understanding science concepts. You can put your own understanding together in ways that your students cannot because they don't see the patterns and features that you do (Bransford, Brown, and Cocking 1999). But you are responsible for helping your students to develop these understandings. The keys to their successful learning are for you to plan experiences that let them grapple with important science concepts and that ensure they make sense of the concepts (Weiss et al. 2003). This is at the heart of the work during the responsive phase. Making student thinking visible is critical. The research into how people in general learn—and in particular how students learn—recommends that teachers determine prior knowledge so they know where to begin instruction (Bransford, Brown, and Cocking 1999; Donovan, Bransford, and Pellegrino 1999). You don't really know what your students are

CHAPTER 1: An Instructional Planning Framework to Address Conceptual Change

thinking unless you purposefully make the effort to find out. Once you identify the nature of any differences between your students' thinking and the science viewpoint, it becomes easier to plan learning experiences (Driver et al. 1994). Identifying the gap in student conceptual understanding helps determine the specific instructional strategies you can use to scaffold your students' learning.

The biggest challenge for science teachers is to build student knowledge and understanding so that students learn the accepted scientific explanations that relate to the hard-to-teach science concepts. However, students build new understandings based on what they already know and believe, which may be inconsistent with the scientific viewpoint. Students see their preconceptions as reasonable and correct, and they may apply them inappropriately to learning situations (Driver et al. 1994). The research into conceptual change reminds us that to change students' ideas that are inconsistent with particular scientific ideas we must present students with new conceptions that appear plausible. Such conceptual change rarely occurs unless students have a chance to engage in inquiries. The new observations and new evidence must get them to think about their thinking in ways that result in the new explanation becoming the more logical or attractive explanation (Strike and Posner 1985). If you do not explicitly address students' everyday conceptions in meaningful ways and ultimately replace their previous conceptions with scientifically more accurate viewpoints, then they will continue to struggle with the conceptual understanding of our hard-to-teach science concepts (NRC 2005).

[2] Through scientific inquiry, students make observations and gather evidence that can change their ideas, deepen their understanding of important scientific principles, and develop important abilities such as reasoning, careful observing, and logical analysis (Minstrell 1989). Inquiry-based learning engages students in the lesson and arouses their interest, promotes teamwork, makes sense out of what is otherwise mystifying, and prepares students to successfully defend findings before an audience of their peers (Layman, Ochoa, and Heikkinen 1996). Students connect their thinking to the inquiry investigations (through hands-on investigations and/or virtual simulations) and create mental models that lead to understanding. Students who use inquiry-based materials understand science concepts more deeply and thoroughly than students who are taught through more traditional methods (Thier 2002).

[3] As revealed in a study of the research, various terms are used for the explanations that students create for themselves as they make sense of scientific phenomena. All of the terms relate to the understanding students have when they arrive in our classrooms. The most prevalent terms are the following:

Preconceptions. This term refers to students' ideas that were formed through life experiences and earlier learning.

Alternative conceptions. This term refers to the variety of ideas that students have that differ from scientific explanations.

Naive conceptions. These are usually incompletely formulated conceptions or simplistic representations of students' conceptual understanding.

Misconceptions. This term refers to students' wrong explanations and errors in thinking.

CHAPTER 1: An Instructional Planning Framework to Address Conceptual Change

Preconceptions, alternative conceptions, and naive conceptions may also include wrong explanations, but the term *preconception* is less negative than *misconceptions*. In this book, *preconception* is used to refer to the student thinking that the teacher hopes to uncover or reveal as he or she implements the Instructional Planning Framework. The term *misconception* is used specifically when referring to misconceptions that have been identified by the considerable research into students' ideas at different grade levels because much of the research uses this term.

Chapter 2

Implementation of the Framework Using the Topic "The Flow of Matter and Energy in Ecosystems"

CHAPTER 2: Implementation of the Framework

> "Science teachers have always used multiple strategies, so we need not make a decision about the one best strategy for teaching science. There isn't one; there are many strategies that can be applied to achieve different outcomes. Science teachers should try to sequence them in coherent and focused ways. This is how inquiry can contribute to the prepared mind."
>
> —Bybee 2006, p. 456

Overview

As a teacher, you know a lot of different strategies to use in the science classroom. However, strategies themselves are not the total solution to learning for your students. You must select those strategies—and activities that target specific science phenomena—that best fit your instructional goals and sequence classroom experiences to support students' conceptual understandings. This chapter helps you put the Instructional Planning Framework (Figure 1.2, p. 8) into practice. The chapter also outlines a ten-stage planning process (Table 2.1) and provides Instructional Tools (pp. 42–44 and 76–126) to help you select strategies that have been shown to develop students' conceptual understandings.

The terms *instructional approach* and *instructional strategy* are used frequently in this chapter. An *approach* is defined as a broad method used in instruction (e.g., reading as a linguistic approach or modeling as a nonlinguistic approach to learning science). *Strategies* are specific ways in which to achieve a specific goal (e.g., use of physical models). As you will discover, although the chapter is grounded in research,[1] it is highly practical in its application to your teaching.

Refer back to the Instructional Planning Framework (Figure 1.2, p. 8). Note again that the framework is made up of two phases: the *predictive phase* and the *responsive phase*. The predictive phase occurs before instruction. As a classroom teacher, you might be called on to participate in that process; however, in many school districts, instructional leaders for all the schools and subjects in the district, rather than individual teachers, complete this work, unpacking the science standards in the process. It is also unlikely for elementary teachers to be involved in the predicative phase because their time for instructional design is so limited—given that they are accountable for the standards for all the subjects they teach. For both of these reasons—the district leaders' responsibility for unpacking the standards and your limited time—only a brief description of the predictive phase (with five stages) is included, but you will learn enough to understand what occurs during this phase.

Table 2.1

Ten-Stage Process for Implementing the Instructional Planning Framework

	Predictive Phase
Stage I	Identify the conceptual target (the learning goal written in terms of "essential understandings" and the related standards and benchmarks).
Stage II	Unpack* the standards and identify the concepts, knowledge, skills, and vocabulary for the content you are teaching.
Stage III	Identify and sequence the subgoals (the learning targets).
Stage IV	Identify criteria for determining student understanding.
Stage V	Identify inquiry and metacognitive goals and strategies.
	Responsive Phase
Stage VI	Research children's misconceptions common to this topic that are documented in the research literature.
Stage VII	Select strategies to identify your students' preconceptions.
Stage VIII	Select strategies to elicit and confront your students' preconceptions.
Stage IX	Select sense-making strategies.
Stage X	Determine formative assessments.

Note: This table will be referred to throughout the book. You might want to flag it to make it easy to find.

*"Unpacking [standards] simply involves identifying the unique elements of information and skill in each ... statement. We have found that subject matter specialists are quite skilled and efficient at doing this task. Consequently, a district need only assemble its ... expert science teachers and curriculum specialists to unpack the science standards ...." (Marzano and Haystead 2008, p. 12)

Most of this chapter, as well as Chapters 4, 5, and 6, focuses on the responsive phase (starting on p. 35) because it is during that phase that you select the instructional strategies you will use during instruction. The life science topic used to demonstrate the framework and Instructional Tools is "the flow of matter and energy in ecosystems." The chapter shows how planning is carried out as if this topic were being taught in the fifth grade. If your school/system addresses this topic at a different grade level, slight modifications can be made (see Chapter 3, where such modifications are explained). Part of the reason that the flow of matter and energy in ecosystems was chosen is that the abstract and conceptual nature of the content make this concept difficult to learn. In addition, it is important to teach this concept well because it builds a foundation for students' study in later grades of photosynthesis and respiration—difficult topics for students and adults alike.

Application of the Predictive Phase to "The Flow of Matter and Energy in Ecosystems"

At the heart of the predictive phase is the identification of learning goals and subgoals and the design of the best instructional sequence to reach these goals. Clarification of the content focus is important for both you and your students. It is equally important that you and your students know and understand the criteria by which the students' understanding will be determined (Stage IV in the predictive phase). Finally, you cannot assume that just because the concept of the flow of matter and energy in ecosystems is covered in your textbook or science kit that this coverage truly addresses the core ideas in science that you want your students to walk away with. So, the question becomes, How might you, during the predictive phase, determine the content focus, the learning sequence, and the criteria for determining student understanding of the topic "the flow of matter and energy in ecosystems"?

Stage I: Identify the Conceptual Target

If your district has not unpacked the standards for you and your fellow teachers, you will need to begin this process with a review of the National Science Education Standards and the Benchmarks for Science Literacy, finding all standards that relate to the content you plan to teach. You will also identify relevant state and district standards. This process can also be used with the emerging Next Generation Science Education Standards (NRC 2010) as they are developed. All of the National Science Education Standards (NSES) (NRC 1996) and Benchmarks for Science Literacy (AAAS 1993, 2009) related to the flow of matter and energy are shown in Table 2.2.

Table 2.2

National Science Education Standards and Benchmarks for Science Literacy Related to "The Flow of Matter and Energy in Ecosystems"*

National Science Education Standards	
Grades K–4	Organisms have basic needs. For example, animals need air, water, and food; plants require air, water, nutrients, and light. (p. 129)
	All animals depend on plants. Some animals eat plants for food. Other animals eat animals that eat the plants. (p. 129)
Grades 5–8	*Populations of organisms can be categorized by the function they serve in an ecosystem. Plants and some microorganisms are producers—they make their own food. All animals, including humans, are consumers, which obtain food by eating other organisms. Decomposers, primarily bacteria and fungi, are consumers that use waste materials and dead organisms for food. Food webs identify the relationships among producers, consumers, and decomposers in an ecosystem.* (p. 157)
	For ecosystems, the major source of energy is sunlight. Energy entering ecosystems as sunlight is transferred by producers into chemical energy through photosynthesis. That energy then passes from organism to organism in food webs. (p. 158)
Benchmarks for Science Literacy	
Grades K–2	Plants and animals both need to take in water, and animals need to take in food. In addition, plants need light. (p. 119)
	Many materials can be recycled and used again, sometimes in different forms. (p. 119)
Grades 3–5	*Almost all kinds of animals' food can be traced back to plants.* (p. 119)
	Some source of "energy" is needed for all organisms to stay alive and grow. (p. 119)
	From food, people obtain energy and materials for body repair and growth.... (p. 136)
	Over the whole earth, organisms are growing, dying, decaying, and new organisms are being produced by the old ones. (p. 119)

*The content in italics is most appropriate for fifth grade, the level for which this unit of study was originally developed.

CHAPTER 2: Implementation of the Framework

Stage II: Unpack the Standards and Identify the Concepts, Knowledge, Skills, and Vocabulary for the Content You Are Teaching

Study the italicized portions of Table 2.2 and translate that information into the *essential understanding* for the unit (as shown in Table 2.3, left-hand column). Then break down each essential understanding into small "chunks" of specific content (Table 2.3, middle column) in order to reduce the gap between your students' current understandings and scientific explanations. Next, identify the most important vocabulary to teach for students' conceptual understandings (Table 2.3, right-hand column). Try not to burden lessons with unnecessary language; it is important that teachers use only language that is essential for scientific literacy and vocabulary that will not require an inordinate amount of time to learn (AAAS 2001a).

Table 2.3

Example of an Essential Understanding With the Necessary Content and Vocabulary for "The Flow of Matter and Energy in Ecosystems" (grades 3–5)

Essential Understanding	Content	Vocabulary
Animals depend on plants for their food source, either by eating them directly or by eating other animals that ate plants. In addition, there are organisms that feed on waste materials and dead organisms. Regardless of where organisms (including humans) get their food, they all need some source of energy to stay alive and grow.	*Learning Target #1:* Food is a substance that provides fuel and building materials for organisms.	Energy Fuel Organisms Substance
	Learning Target #2: Plants need sunlight, air, and water to make food and to grow.	Plants Growth
	Learning Target #3: Animals get food either by eating plants directly or by eating animals that ate plants.	Animals Consumers Food Eating Plants Producers
	Learning Target #4: Some organisms use waste materials and dead organisms for food, decomposing both and returning matter to the soil.	Death Decay Decomposers Growth Waste Materials

Stage III: Identify and Sequence the Subgoals

Now sequence the chunks of content. These become your *learning targets*. Make sure to build on ideas taught in earlier grades and during the school year; these ideas become the *prerequisite knowledge/skills for the lesson*. First teach ideas that are foundational to understanding the ideas that will follow. Figure 2.1 shows the learning sequence developed for a unit on the flow of matter and energy in ecosystems.

Figure 2.1

Learning Targets and Sequence for "The Flow of Matter and Energy in Ecosystems"

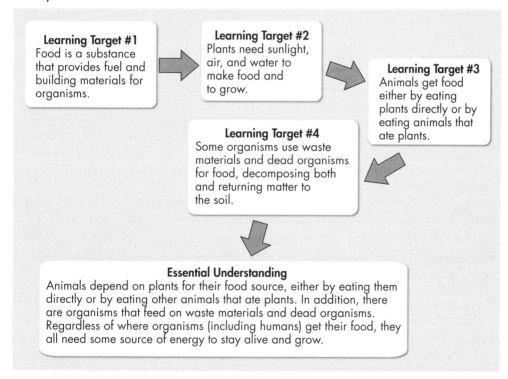

Stage IV: Identify Criteria for Determining Student Understanding

Establish criteria by which to measure student understanding, with one criterion for each learning target. Criteria are what you should look for when you examine student products and performances to determine student success or acceptability of work. In other words, "[criteria are] the qualities that must be met for work to measure up to a standard" (McTighe and Wiggins 1999, p. 275). It is very important that you identify these criteria *before* you think about activities or even specific performance tasks. Consider using the "backwards design" process as described in *Understanding by*

Design (Wiggins and McTighe 1998) if your school district has provided professional development on the implementation of this model.

You will eventually use two basic forms of student assessment: summative and formative. Summative assessments are cumulative assessments that try to capture what a student has learned and can be used for grading, placement, promotion, or accountability (NRC 2001). However, this type of assessment is not what is intended in Stage IV of the predictive process. Instead, at this stage you should establish criteria by which students' understandings are formatively assessed so that you can use students' current understandings to modify your instruction. The form of assessment you use depends on your context. It is ideal to focus on big ideas and conceptual understanding, but if assessments in your district focus instead on facts and vocabulary, you'll need to teach both the big ideas AND facts and vocabulary. You'll have to use both formative and summative assessment to ensure that your students understand both the big ideas and the facts and vocabulary.

As stated earlier, most of you are not responsible for planning in the predictive phase, especially these first four stages. However, there is a lot to learn from going through the process yourself, especially because it provides a deep look into the content you teach. If you decide you want to tackle this work for a science topic, it is best done with a team or your professional learning community. Planning hints and resources for this undertaking are listed in Figure 2.2.

Figure 2.2

Predictive Phase: Hints and Resources for Planning in Stages I–IV

Hint: It is important that you understand the unit content. Consider these resources.

Resources:
- *Science Curriculum Topic Study: Bridging the Gap Between Standards and Practice* (Keeley 2005) provides a systematic strategy that begins with topics and connects standards and research with curriculum, instruction, and assessment.
- *Science Matters: Achieving Science Literacy* (Hazen and Trefil 2009) provides accessible explanations of core science concepts.
- Bill Robertson's *Stop Faking It!* series, published by NSTA Press (*www.nsta.org/store/search.aspx?action=quicksearch&text=stop%20faking%20it%20series*) explains science basics for a variety of topics and provides easy-to-follow activities that help teachers learn the fundamental concepts.
- National Science Teachers Association (NSTA) Learning Center resources (*http://learningcenter.nsta.org/?lid=lnavhp*) include various texts and online resources that target your particular learning needs.

Figure 2.2 *(continued)*

Hint: Working with standards—not the textbook—is the starting place for instruction. Review your district and state standards as well as the national science standards.

Resources:
- The National Science Education Standards are at *www.nap.edu/openbook.php?record_id=4962*.
- Go to *www.project2061.org/publications/bsl/online/index.php* for the Benchmarks for Science Literacy.

Hint: Unpack your standards into small chunks so that the gap between your students' understandings and the scientific explanations are diminished. Many words used in the standards are actually concepts for your students to understand.

Resources:
- *Making Sense of Secondary Science: Research into Children's Ideas* (Driver et al. 1994) summarizes research about misconceptions for students from as early as five years old through high school and beyond.
- *Benchmarks for Science Literacy* (AAAS 1993, 2009)
- *National Science Education Standards* (NRC 1996)

Hint: Some concepts and vocabulary in the standards are mandatory if students are to develop essential understandings. Others are not. These resources should help you separate the two.

Resources:
- *Designs for Science Literacy* (AAAS 2001a) includes a process for and examples of how to "prune" the content in the standards.
- Chapter 2 of *Designing Effective Science Instruction: What Works in Science Classrooms* (Tweed 2009) includes a nice description, with helpful examples, of how to distinguish between mandatory and nonmandatory concepts and vocabulary.

Hint: It is important to chunk the essential understandings into subgoals (learning targets) so that the gap between the targeted content and your students' understandings is not too great.

Resources:
- Study the strand maps in the two volumes of the *Atlas for Science Literacy* (*http://strandmaps.nsdl.org*).
- A simple Google search (e.g., "learning progressions" + your topic + 3–5 classrooms) should yield current research about logical learning progressions.
- Review instructional materials available in your school and school district and identify activities that connect to the learning targets.

(continued)

Figure 2.2 *(continued)*

> **Hint:** Carefully consider prerequisites for a unit or lesson and decide what you will do if some students have not learned the prerequisites. Cross-grade-level meetings in your school should clarify learning progressions and possible integrated approaches to instruction that blend life, physical, and Earth and space science concepts.

Resources:
- Your own district standards should be your initial guide in this process.
- Once again, the strand maps from the two volumes of the *Atlas for Science Literacy* should help.

> **Hint:** Develop criteria using the various facets of understanding used in "backwards design." The work by Grant Wiggins and Jay McTighe includes several valuable resources that make it clear that essential understandings must come first in the design process.

Resources:
- *Understanding by Design* (Wiggins and McTighe 1998)
- *The Understanding by Design Handbook* (McTighe and Wiggins 1999)
- *The Understanding by Design Professional Development Workbook* (McTighe and Wiggins 2004)

> **Hint:** The criteria you develop can be used to create formative and summative assessments as well as rubrics. In addition to the work by Wiggins and McTighe listed above, some excellent science-specific resources follow.

Resources:
- *Assessing Student Understanding in Science.* 2nd ed. (Enger and Yager 2009)
- *Assessment for Learning: Putting It Into Practice* (Black et al. 2003)
- *Assessment in Science: Practical Experiences and Education Research* (McMahon et al., eds. 2006)
- *Everyday Assessment in the Science Classroom* (Atkin and Coffey, eds. 2003)

Stage V: Identify Inquiry and Metacognitive Goals and Strategies

Students are more likely to learn when you use strategies to increase their motivation and engage them with content. Because research has found that *inquiry* and *metacognition* strategies contribute to student motivation and engagement, it is wise to have at least one inquiry goal and one metacognitive goal in each instructional unit. Students' exposure to various inquiry and metacognitive skills during the year enhances not only those skills but content understandings as well. This section of the books describes how to select inquiry and metacognitive strategies.

Inquiry

Inquiry learning requires students to grapple with ideas and pose explanations. Inquiry is not required for every lesson, but it is appropriate for hard-to-teach-and-

CHAPTER 2: Implementation of the Framework

learn concepts because it calls for students to engage deeply with the content, promoting conceptual change. With the more traditional, "recipe" investigations, students may misinterpret the experience or miss the core ideas.

Inquiry means different things to different people, but this book focuses on the five "essential features" of inquiry outlined in *Inquiry and the National Science Education Standards: A Guide for Teaching and Learning* (NRC 2000, p. 29). This framework (Figure 2.3) shows a continuum of variations that give you and your students varying amounts of control over the inquiry process. You might be familiar with the terms *open, guided,* and

Figure 2.3

The Five Essential Features of Classroom Inquiry and Their Variations

Essential Feature	Variations — More ← Amount of Learner Self-Direction → Less / Less ← Amount of Direction From Teacher or Manual → More			
1. Learner engages in scientifically oriented questions	Learner poses a question	Learner selects among questions, poses new questions	Learner sharpens or clarifies question provided by the teacher, materials, or other sources	Learner engages in question provided by teacher, materials, or other sources
2. Learner gives priority to **evidence** in responding to questions	Learner determines what constitutes evidence and collects it	Learner directed to collect certain data	Learner given data and asked to analyze	Learner given data and told how to analyze
3. Learner formulates **explanations** from evidence	Learner formulates explanation after summarizing evidence	Learner guided in process of formulating explanations from evidence	Learner given possible ways to use evidence to formulate explanation	Learner provided with evidence
4. Learner connects explanations to scientific knowledge	Learner independently examines other resources and forms the links to explanations	Learner directed toward areas and sources of scientific knowledge	Learner given possible connections	
5. Learner communicates and justifies explanations	Learner forms reasonable and logical argument to communicate explanations	Learner coached in development of communication	Learner provided broad guidelines to use to sharpen communication	Learner given steps and procedures for communication

Source: National Research Council (NRC). 2000. *Inquiry and the national science education standards: a guide for teaching and learning.* Washington, DC: National Academy Press, p. 29. Reprinted with permission.

Hard-to-Teach Science Concepts

CHAPTER 2: Implementation of the Framework

structured inquiry. In Figure 2.3 the variations closer to the left—in which the learner is responsible for each of the essential features—are what you might know as open inquiry. The farther to the right you move, the more guided the inquiry. The far right-hand column shows examples of structured inquiry. The three steps that follow outline how you can identify your inquiry focus.

Step 1. Determine the inquiry approach to include in your instructional plan. With guided inquiry, students develop specific content knowledge because you determine the learning goals. In contrast, open-ended inquiry promotes students' understandings of inquiry and the nature of science. Neither guided nor open-ended inquiry is "correct"; each is used for different purposes depending on your students' needs, the time in the school year, where the lesson is placed in a unit of study, and the content focus.

You may want to start your year with more guided inquiry and then, during the year, help your students to develop their use of the five "essential features" in Figure 2.3, especially if they have never experienced inquiry. You can later provide more open-ended experiences as their abilities and understandings improve. Or you may start a unit with more guided inquiry and allow an open-inquiry experience later in the unit when you want your students to apply what they have learned. Consider when to best use more guided or open inquiry based on the specific content and the experiences of your student group.

How does inquiry apply to our instructional unit on the flow of matter and energy in ecosystems? Many teachers explore ecosystem-related topics early in the school year for a variety of reasons, including weather conditions. Because the study of food chains begs for outdoor observations and investigations, called field experiments, this unit is developed for fall implementation and will include student tasks in the field (fieldwork). Some guidance in inquiry will be required, however, as this might be the first real inquiry opportunity for your students. Thus, the lessons that appear in Table 2.15, page 60, the Planning Template for the Responsive Phase, include inquiry options from farther to the right in Figure 2.3.

Step 2. Select the specific feature of inquiry on which to focus. The *Atlas of Science Literacy*, Vol. 1 (AAAS 2001b) will help you in that decision. Take a look at a portion of a strand map from the atlas in Figure 2.4. (*Note:* The white-shaded boxes are abilities and the gray-shaded boxes are understandings.) Notice that at the K–2 level, there is a focus on asking questions and making observations, while at the 3–5 level the focus shifts to different abilities that include distinguishing observations from speculations, providing reasons (evidence) for what students believe happened, and keeping a careful record (notebook) of the observations and reasoning.

As mentioned earlier, every lesson need not include inquiry. But over the course of the school year, you should work to develop your students' understanding of and abilities in each of the essential features. Study Instructional Tool 2.1, Teaching the Five Essential Features of Inquiry (pp. 76–82), which describes research related to each

Figure 2.4
Partial Strand Map for "Scientific Inquiry: Evidence and Reasoning in Inquiry"

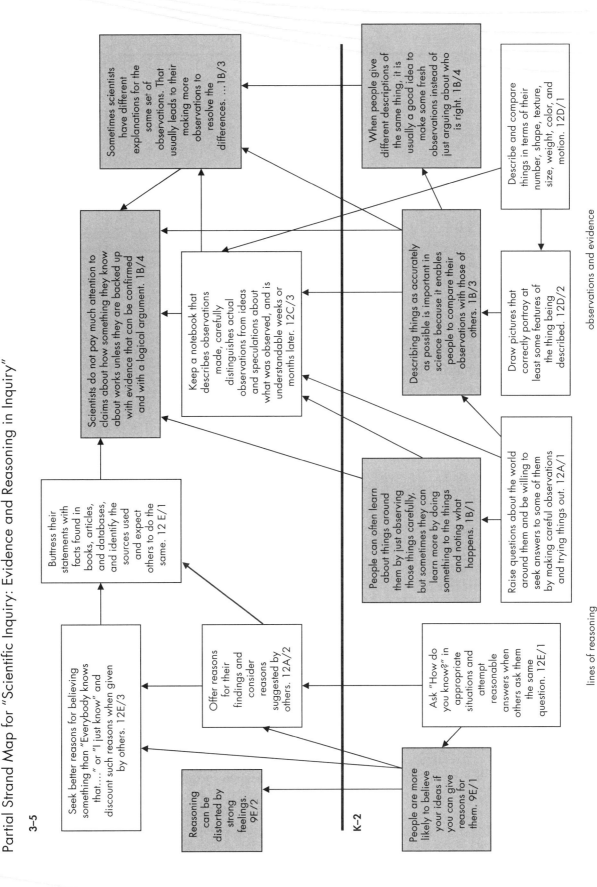

Source: American Association for the Advancement of Science (AAAS). 2001b. *Atlas of science literacy.* Vol. 1. ("Scientific Inquiry: Evidence and Reading in Inquiry" map, p. 17.) Washington, DC: AAAS.

of these features, outlines classroom applications, gives examples of implementation, and highlights technology applications. Read through each of the identified strategies and determine which best fits your needs.

Keep in mind that the unit under discussion in this chapter—the flow of matter and energy in ecosystems—is intended to be used in the fifth grade near the beginning of the school year. If you teach this topic as early as third grade, you might want to focus more on the students' abilities to distinguish observations from speculations. But by fifth grade, you want your students to become more competent with gathering appropriate evidence—that is, knowing that evidence is required to respond to questions they are asked by you or their peers.

Eventually, you also want them to develop the abilities to formulate explanations based on this evidence, but since it is early in the school year, this lesson focuses on the second essential feature of inquiry (see Figure 2.3): "Learner gives priority to evidence in responding to questions" (in other words, students need to learn to cite evidence when they are answering questions from their peers or the teacher).

Step 3. Establish the criterion to determine your students' understanding of a target inquiry goal. The criterion developed for this particular life science unit is "Use appropriate methods to provide evidence required to support a claim and build an explanation."

Metacognition

Cognition is the thought process—processing information and applying knowledge—and *metacognition* is thinking about thinking—reflecting on your own thinking. Metacognition requires that your students are aware of and have control over their learning. You should address metacognitive goals (in addition to cognitive goals) in each lesson because *teaching metacognitive strategies is second only to classroom management in influencing student learning* (Wang, Haertel, and Walberg 1993/94).

Thinking is invisible, but there are ways that you can make it visible to your students, helping them to become more metacognitively aware and to see school as more about exploring ideas than memorizing content. Table 2.4 describes three metacognitive approaches (Marzano 1992) that, when used in your classroom, will increase the likelihood that your students will develop expertise in these areas at the same time that they are learning science content. (*Note:* These three approaches are explored at greater length in Instructional Tool 2.2, p. 83.)

Step 1. First determine (from Table 2.4) the metacognitive approach for your lesson, keeping in mind that during the year you want to develop your students' abilities in each area. The sequence is not critical, but you might want to work on "self-regulated thinking" early in the school year. This will quickly establish student-centered instruction in your classroom. Also, self-regulated thinking is an appropriate metacognitive focus in the flow of matter and energy in ecosystems unit because it can help students plan and organize their fieldwork.

Table 2.4

Three Metacognitive Approaches

Critical Thinking and Learning	"Critical thinking and learning include being accurate and seeking accuracy, being clear and seeking clarity, being open-minded, restraining impulsivity, taking a position when the situation warrants it, and being sensitive to others' feelings and level of knowledge"(Marzano 1992, pp. 133–134).
Creative Thinking and Learning	"Creative thinking and learning include engaging intensely in tasks even when answers or solutions are not immediately apparent, pushing the limits of your knowledge and abilities, generating, trusting, and maintaining your own standards of evaluation, and generating new ways of viewing a situation outside the boundaries of standard conventions" (Marzano 1992, p. 134).
Self-Regulated Thinking	Self-regulated thinking includes "being aware of your own thinking, planning, being aware of necessary resources, being sensitive to feedback, and evaluating the effectiveness of your actions" (Marzano 1992, p. 133).

Step 2. Refer to Instructional Tool 2.2, Three Strategies That Support Metacognition (p. 83), to learn more about your selected approach—in this case, Self-Regulated Thinking, on page 84. Consider the strengths of each strategy and determine which best fits your planned instruction. For the flow of matter and energy in ecosystems unit, the strategies Plan and Self-Regulate and Self-Evaluate could be important to students as they manage their fieldwork. Group discussion is essential to research and experimentation, so Talk About Thinking is also appropriate.

The Plan and Self-Regulate strategy was chosen for this lesson because it most clearly aligns with the learning targets and the plans for fieldwork. Note the description of this strategy (on p. 86): "This strategy involves estimating time requirements, organizing materials, scheduling procedures needed to complete an activity, and developing evaluation criteria." This sounds perfect for the instructional plan that we are building in this chapter. Research indicates that graphic organizers, such as the one in Table 2.5, page 32, can help students a great deal during this process. You can give a blank version of this organizer to students and require that they complete it as they organize for their fieldwork. Check their plans before the fieldwork begins.

CHAPTER 2: Implementation of the Framework

Table 2.5

Work-Plan Organizer With Fieldwork Example

Instructions: As you plan for your fieldwork, complete the following organizer. Define the tasks you must complete to get ready for work. Then determine the materials needed for each task, the team member(s) responsible for doing each task, the procedures you will follow, and the deadline to complete the work.

What Is Our Task?	What Materials Do We Need?	What Are the Procedures?	Who Will Do the Work?	By When Will the Work Be Done?
Example: Gather soil samples.	4 plastic bags 2 trowels	1. Identify four spots to collect soil. 2. Go to each spot and dig two inches below the surface. 3. At each spot, fill a trowel with soil so that we have samples from each of the four spots.	Sara and Michael	The first day of our fieldwork

Step 3. Develop the criterion to determine your students' understanding of this metacognitive goal. The metacognitive criterion developed by the authors for flow of matter and energy in ecosystems is, "Carefully plan and implement group research, monitoring personal and group progress." Before moving on, review the hints and resources for Stage V (Figure 2.5).

Figure 2.5

Predictive Phase: Hints and Resources for Planning in Stage V

Hint: Inquiry is best learned in conjunction with content. You must carefully select strategies that align with your students' needs and the content you are teaching as well as consider the best time of the school year to introduce inquiry strategies.

Resources:
- *Inquire Within: Implementing Inquiry-Based Science Standards in Grades 3–8* (Llewellyn 2007) provides insights and strategies to implement inquiry-based instruction as well as practical examples of how to modify lessons to make them more inquiry-based.
- *Inquiry and the National Science Education Standards: A Guide for Teaching and Learning* (NRC 2000)

Hint: Metacognitive abilities are best improved when taught in conjunction with content learning (cognitive focus). Carefully select strategies that align well with your students' needs, the time of year, and the particular content you are teaching.

Resources:
- The three metacognitive approaches used in this book are drawn from Marzano's (1992) work *A Different Kind of Classroom: Teaching with Dimensions of Learning*. This is a wonderful resource to extend your learning about ways to improve your students' metacognitive abilities.
- The Visible Thinking website (*www.pz.harvard.edu/vt*) provides excellent tools and ideas that promote metacognition.
- Other good critical- and-creative thinking websites include the following:
  - www.engin.umich.edu/~problemsolving/strategy/crit-n-creat.htm
  - http://members.optusnet.com.au/~charles57/Creative/Techniques/index.html
  - www.mindtools.com/pages/main/newMN_CT.htm

---

* * * * * *

The process for the predictive phase is now complete. A summary of the work we have done is shown in the Planning Template (Table 2.6, p. 34). A blank version of this template appears as Appendix A. As you can see, about two-thirds of the way down on Table 2.6, the instructional plans for each of the learning targets will be completed during the responsive phase, which we tackle next.

CHAPTER 2: Implementation of the Framework

Table 2.6

Predictive Phase: Planning Template for "The Flow of Matter and Energy in Ecosystems"

	Lesson Topic: "The Flow of Matter and Energy in Ecosystems"
Essential Understandings	Animals depend on plants for their food source, either by eating them directly or by eating other animals that ate plants. In addition, there are organisms that feed on waste materials and dead organisms. Regardless of where organisms (including humans) get their food, they all need some source of energy to stay alive and grow.
Knowledge Required From Previous Instruction	• Differences between plant and animal (K–2) • Basic needs of plants and animals (K–2) • Differences between living and dead (3–5) • Air is a substance that surrounds us and takes up space (3–5) • States of matter and the role of heat in changing states (3–5) • The role of scale in understanding food chains (3–5)
Knowledge and Skills to Be Learned	**Concepts:** See the Learning Targets on the next page. **Vocabulary:** *energy, fuel, organisms, substance, animals, consumers, food, eating, plants, producers, death, decay, decomposers, growth, waste materials, life, food chain* **Skills:** Estimate time requirements, determine materials needs, and assign tasks for group work; determine procedures to gather data aligned with inquiry question; formulate explanations based on evidence; use tools and measurement devices that are appropriate to designed experiments
Criteria to Demonstrate Understanding	• Provide a thorough explanation of how food is essential to an organism's growth and life. • Conduct an experiment to determine the requirements of plant growth and describe the flow of matter that occurs. • Carefully represent the flow of matter from plants through the various animals in a food chain. • Predict what would happen if there were no decomposers and support the prediction with understandings about matter. • Use appropriate methods to provide evidence required to support a claim and build an explanation (inquiry focus). • Carefully plan and implement group research, monitoring personal and group progress (metacognitive focus).

Table 2.6 (continued)

\multicolumn{2}{l}{Learning Targets and Instructional Plans for "The Flow of Matter and Energy in Ecosystems"}
Target #1	Food is a substance that provides fuel and building materials for organisms.
	Instructional Plan for Target #1: Will be completed during the responsive phase (see Table 2.15, p. 60).
Target #2	Plants need sunlight, air, and water to make food and to grow.
	Instructional Plan for Target #2: Will be completed during the responsive phase (see Table 2.15).
Target #3	Animals get food either by eating plants directly or eating animals that ate plants.
	Instructional Plan for Target #3: Will be completed during the responsive phase (see Table 2.15).
Target #4	Some organisms use waste materials and dead organisms for food, decomposing both and returning matter to the soil.
	Instructional Plan for Target #4: Will be completed during the responsive phase (see Table 2.15).

Application of the Responsive Phase to "The Flow of Matter and Energy in Ecosystems"

During the predictive phase, you determined the content focus, established a learning sequence, developed criteria to demonstrate understanding, and identified inquiry and metacognitive goals. In the responsive phase, you will decide what the learning experiences will be in your classroom, specifically selecting strategies that uncover students' ideas and promote conceptual understanding. The responsive phase is so named because your plan may change *in response to* your students' learning needs.

First, select activities from your textbook, NSTA resources, the internet, and other sources of teaching ideas that address the content of the learning targets regarding the flow of matter and energy in ecosystems. A good curriculum will sequence activities appropriately to build students' content understanding. But if the curriculum you have to work with does not do this, you need to look elsewhere for activities that both represent the content and link to the targets. Once you find the activities, you may need to modify them and/or sequence them appropriately.

CHAPTER 2: Implementation of the Framework

Of course, even the best activities do not ensure conceptual understanding. That is why the instructional strategies you choose are so important. The plan that results from the responsive phase is your best prediction, based on research, of what will work in your classroom for a given topic.

Stage VI: Research Children's Misconceptions Common to This Topic That Are Documented in the Research Literature

During this stage, you compile and study the misconceptions that research has found to be common among children in relation to the flow of matter and energy in ecosystems. Consider the implications of these misconceptions for instruction. See Figure 2.6 for some hints and resources that will help in this process. Of these resources, the most readily accessible (and free!) are the misconceptions provided at the NSDL website. (*Note:* The authors have begun the search for misconceptions for you, and their findings—together with instructional ideas to confront the misconceptions—are summarized in Table 2.7, pp. 38–40.)

Figure 2.6

Responsive Phase: Hints and Resources for Planning in Stage VI

Hint: Research helps us determine common misconceptions that our students have. Adults often have the same misconceptions. Your identification of these misconceptions will help you better understand the content and design lessons that target these misconceptions in your students.

Resources:
- The National Science Digital Library (NSDL) has science literacy maps and resources online. These present the strand maps developed by Project 2061 (AAAS 2001b, 2007). If you have these resources in book form, you can use them. If not, visit the NSDL site. Go to *http://strandmaps.nsdl.org* and select a topic. For the work we're doing in this chapter, you would select "The Living Environment" and then "Flow of Energy and Matter." In the upper left-hand corner, you will find a link that says "View Student Misconceptions."
- An excellent resource is *Making Sense of Secondary Science: Research into Children's Ideas* (Driver et al. 1994). This book provides a very nice summary of various science topics and the related research. Note that although the word *secondary* is used in the title, this resource outlines misconceptions commonly held by students in grades 3–5 and above.
- It's worth trying to google for misconceptions. For example, you can search for "food webs and misconceptions," "matter and misconceptions," or "ecosystems and misconceptions."
- University research databases provide access to the most recent research.

Figure 2.6 *(continued)*

> **Hint:** Resources exist that will help you directly target specific misconceptions and provide good instructional ideas as you begin to plan your lessons.

Resources:
- *Making Sense of Secondary Science: Research into Children's Ideas* (Driver et al. 1994) is the classic work on this topic.
- The *Uncovering Student Ideas in Science* series by Page Keeley and her colleagues includes numerous probes that help you determine and explore your students' ideas about many science concepts. This series is available through NSTA Press (*www.nsta.org/publications/press/uncovering.aspx*).

CHAPTER 2: Implementation of the Framework

Table 2.7

Student Misconceptions on "The Flow of Matter and Energy in Ecosystems" (Results of Literature Search)

Concept	Misconceptions	Instructional Ideas
Food is a substance that provides fuel and building materials for organisms.	It is critical for students to understand the scientific meaning of *food*. They tend to give nonfunctional explanations for the need for food. They know food is necessary for life and helps them grow, stay healthy, and be active, but the meaning they give to the word *food* varies in different contexts. They say food is needed to stay alive but do not describe the role of food in metabolism. They think food is needed for growth but do not understand that it is a source of matter for growth. They have limited understandings about the transformation of food as it is made a part of the growing body (AAAS 1993; Driver et al. 1994).	Start with students' ideas about what they think food is (e.g., creative writing, posters) and share those ideas (Keeley and Tugel 2009; Leeds National Curriculum Science Support Project [LNCSSP] 1992b). Compare how their definitions of food differ from the scientific definition. Provide examples so they can see the difference and begin to understand the idea that food provides energy and material for growth. Be sure to provide examples of all kinds of organisms, including plants (Keeley and Tugel 2009).
	Many 11- and 12-year-olds think that food vanishes and is converted into "energy" or "goodness" (Driver et al. 1994). Children often think materials appear and disappear (LNCSSP 1992a); for example, food and air "disappear" once taken into the body (Driver et al. 1994). They have limited understandings about the transformation of food as it is made a part of the growing body (AAAS 1993; Driver et al. 1994).	Since young children hold nonconservation ideas like vanishing food, challenge them with questions such as, "How does what we eat become skin, bone, and other parts of the body?" Using the word *biomass* helps to build the concept that what is eaten by animals or made by plants becomes part of the organism. Students should also be asked to consider questions about the origin of the parts of an organism (LNCSSP 1992b).
	Children who are 8 and 9 years old think that all moving things are alive and only those things. By age 9–11, this definition shifts to things that move by themselves (including rivers and the Sun). By age 11, most include as living either animals alone or animals and plants (Piaget 1929). Many studies show that even up to age 15, students still make errors categorizing living and nonliving things. Very few understand *living* in a biological way. They often overextend the definition, a result of using some but not all characteristics of living things (e.g., using the criterion of movement to categorize clouds as living). They are more often correct when categorizing animals than when categorizing plants. The most common criteria used to define living were movement in animals and growth and development in plants (Driver et al. 1994).	Let students observe a variety of organisms and generate characteristics of living things. Use a variety of organisms that include plants and animals, parts of organisms (e.g., carrot tops in water), seeds, etc. Also have them observe less easily categorized objects to determine whether or not they are living. Include items like rocks, fossils, and dried pieces of sponge and determine ways to find out if they are living, were once living, or were never alive. By third grade also include microorganisms. Pay special attention to phrases used that imply nonliving things do what living things do (e.g., waves "grow," wind "whistles"). Help students learn the seven life processes (movement, respiration, reaction to stimuli, growth, reproduction, elimination of wastes, and nutrition) (Keeley, Eberle, and Farrin 2005).

Table 2.7 *(continued)*

Concept	Misconceptions	Instructional Ideas
Plants need sunlight, air, and water to make food and to grow.	At this age, students translate their ideas of eating and growth to organisms, so they consider anything taken into the plant as food (LNCSSP 1992b).	Students can become familiar with common plants by using drawings or photographs. These visual representations can show the plant parts that take in water and sunlight (not *how* they take in light). Ask students to identify sources of the materials and energy needed by plants (WestEd and CCSSO 2007).
	Students often think that plants get their food from soil through their roots and that water, minerals, fertilizers, carbon dioxide, and sunlight are plant food (Driver et al. 1994). They often think of water and air as what a plant drinks and breathes (LNCSSP 1992b). Most students at this age think that plants do not use air or that plants and animals use air in opposite ways (Driver et al. 1994).	Students should be aware of the parts of a food chain and that plants need sunlight. But it is hard for them to understand that plants make their own food, and this instruction should occur in the middle grades (AAAS 1993). The use of the word *photosynthesis* should be avoided until an older age when students can attach meaning to it. Elicit and discuss students' ideas about how they think plants get food. It might also be helpful to use the analogy of a factory, talking about "raw materials" taken from the environment to make the product, "food." These analogies can be developed through creative writing, posters, group discussion, and/or drawings (LNCSSP 1992b).
Animals get food either by eating plants directly or by eating animals that ate plants.	Upper elementary students translate their ideas of "eating" and "growth" to organisms, so they consider anything taken into the plant or animal as food (LNCSSP 1992b). Primary-level students think of individual organisms that people keep and care for, but "pupils aged 7–11 extend their thinking to wild organisms as individuals, although some may think that these are fed and cared for by people" (Driver et al. 1994, p. 59).	Students can become familiar with common animals (including vertebrates and insects) by using drawings or photographs. These visual representations can show animals' external organs that take in air, water, and food. Ask them to identify the sources of the materials and energy needed by animals (WestEd and CCSSO 2007).
	It is uncommon for students to relate their ideas about feeding and energy to organisms' interactions (i.e., food chains and food webs). Older elementary students who know that animals can't live without plants still do not relate this to the inability of animals to make their own food. Students' understandings of ecological relationships depend on their understanding of *producer* and *consumer*. And this depends on their concepts of *plant* and *animal* (Driver et al. 1994). Preteens (11- and 12-year-olds) see predation as an "eating event" for the benefit of the eater. They regard food that is eaten as part of the food chain as being different from food incorporated into the eater's body (Smith and Anderson 1986).	Study of food chains can begin at this age, but you should not attempt to teach flow of energy through organisms or storage of energy in molecules. Students should be aware of the parts of a food chain and that animals eat plants or other animals (AAAS 1993). To help students move from their knowledge of "eating events," you need to help them understand that the body material of one organism (the eaten) becomes the body material of another (the eater) (LNCSSP 1992a). Using the matter and energy process tool (Figure 2.12, p. 14) in conjunction with food chains may help in the process. (The tool was developed by Mohan and Anderson [2009] with a team from Michigan State University.)

(continued)

Table 2.7 *(continued)*

Concept	Misconceptions	Instructional Ideas
Some organisms use waste materials and dead organisms for food, decomposing both and returning matter to the soil.	Young children think dead things disappear. Even as they get older they think some or all of the matter from the organism disappears. They do not understand conservation of matter (Leach et al. 1992).	It is inappropriate to teach about geochemical cycles, but awareness of recycling lays a foundation for future learning about the transfer and conservation of matter (AAAS 1993).
	Young children do not understand that material from the dead organism becomes part of the environment or that microbes are involved in the process. They think that insects are involved, and it is common for them to think that bugs or germs eat some of the rotted matter. They do not recognize the role of microbes in the decay process (Driver et al. 1994).	Give students opportunities to observe changes during decomposition and link those changes to a living organism, but be certain to expose students to more than one type of organism involved in decay so they do not think too narrowly about the types of organisms involved in the process. You might consider using the "Rotting Apple" probe (Keeley, Eberle, and Dorsey 2008) to connect the ideas of decay and food. The authors of the probe suggest that you specifically ask students what they think food is and if a rotten apple or decaying animal serves as food for any organisms.
Organisms grow and eventually die.	Before age 9, students lack the concepts that are required to understand the biological notions of growth. Even high school students tend to use the word *growth* as related to size and "getting bigger." From an early age, students think eating or absorbing material is essential for growth. But they often fail to understand that the substances taken in are the material basis for growth and, once in the body and transformed, make the body bigger (Driver et al 1994).	"At this grade level, growth is understood at the macroscopic level and connected to the needs of organisms, such as food being a requirement for growth. Students can observe and measure an organism's growth, but a cellular and molecular explanation is not expected until middle school" (Keeley, Eberle, and Dorsey 2008, p. 127).
	A majority of 11-year-old children give identical meaning to *destruction* and *dying*. They begin to understand death as a biological process with the body ceasing to function by about age 9 or 10, but they think of it in human terms, not always seeing it as final and inevitable. Teaching doesn't change their ideas about death or decomposition. Most think of death in terms of animals, but not of all living organisms (Driver et al. 1994).	Emphasis should be placed on observing various plants and animals and considering what it is they need to stay alive (AAAS 1993). At the grades 3–5 level, it is inappropriate to focus on the function of cells in sustaining life or all the activities going on in cells. This focus should be reserved for the middle grades (AAAS 1993).

Instructional Strategy Selection Tool: Identifying Strategies for the Responsive Phase

Notice that Stages VII–X (see Table 2.1 for a list of the stages) are about strategy selection. You need to find strategies that will that help *identify, elicit,* and *confront* your students' preconceptions, as well as strategies that help them *make sense* of those experiences. Coupling these strategies with effective activities (which address the targeted phenomenon and link with the learning targets) helps ensure that selected activities will build conceptual understanding.

This section of the chapter is devoted to Instructional Tool 2.3, the Instructional Strategy Selection Tool (pp. 42–44), used to identify potential strategies for use in the responsive phase. Specific strategies in Instructional Tool 2.3 are listed according to one of two approaches—linguistic representations (Instructional Tools 2.5–2.6, pp. 98–106) and nonlinguistic representations (Instructional Tools 2.7–2.10, pp. 107–126)—and then categories within each approach. (All of the Instructional Tools are at the end of this chapter *except* for Instructional Tool 2.3, which has been pulled out and discussed here because it is the basis for the other Instructional Tools.)

CHAPTER 2: Implementation of the Framework

Instructional Tool 2.3 (See Notes Below)

Instructional Strategy Selection Tool

		Framework Components							
		Identifying Preconceptions	Eliciting and Confronting Preconceptions		Sense Making			Demonstrating Understanding	
		Bringing (by the teacher) students' preconceptions to the surface and determining prior knowledge	Explicitly eliciting preconceptions	Confronting preconceptions	Perceiving, interpreting, and organizing information	Connecting information	Retrieving, extending, and applying information; using knowledge in relevant ways	Formative assessment	Peer and self-assessment
Linguistic Representations of Knowledge									
Approaches	**Strategies**								
Writing to Learn (Instructional Tool 2.4, p. 90)	Learning Logs	√	♥		♥	√	√	√	♥
	Probes		♥	√				♥	♥
	Science Notebooks		√	√	♥	♥	√	♥	√
	Scientific Explanations		√	♥	♥	√	√	♥	√
	Science Writing Heuristic		♥	♥	♥	♥	√	♥	√

Notes: This Instructional Tool will be referred to throughout the book. You might want to flag it to make it easy to find. A check mark (√) means that that strategy will promote student understanding, if effectively implemented. A heart symbol (♥) means that that strategy is very strongly supported by research (and is also a favorite of the authors!).

Instructional Tool 2.3 *(continued)*

Approaches	Strategies	Identifying Preconceptions	Eliciting and Confronting Preconceptions		Sense Making			Demonstrating Understanding	
Reading to Learn (Instructional Tool 2.5, p. 98)	Vocabulary Development Strategies		√		♥	♥		√	
	Informational Text Strategies	♥	√	♥	♥	♥	√	√	
	Reflection Strategies		√	√	√	♥	♥	√	♥
Speaking to Learn (Instructional Tool 2.6, p. 103)	Large- and Small-Group Discourse (including questioning)	√	♥	♥	√	√	√	♥	♥
	Student Questioning		♥	♥				√	♥

Nonlinguistic Representations of Knowledge

Approaches	Strategies								
Models (Instructional Tool 2.7, p. 107)	Mathematical Models				♥	♥		√	
	Physical Models			√	♥	♥	√	√	♥
	Verbal Models: Analogies				♥	√	♥	√	√
	Verbal Models: Metaphors				♥	√	♥	√	√
	Visual Models		♥	♥	√	♥	√	√	√
	Dynamic Models		♥	√	√	♥	♥	√	√

Note: A check mark (√) means that that strategy will promote student understanding, if effectively implemented. A heart symbol (♥) means that that strategy is very strongly supported by research (and is also a favorite of the authors!).

(continued)

Instructional Tool 2.3 *(continued)*

Approaches	Strategies	Identifying Preconceptions	Eliciting and Confronting Preconceptions		Sense Making			Demonstrating Understanding	
Visual Tools (Instructional Tool 2.8, p. 114)	Brainstorming Webs	♥	√		√	♥		√	
	Task-Specific Organizers		√		♥	♥	√	√	♥
	Thinking-Process Maps	√	♥	♥	♥	♥	√	♥	♥
Drawing Out Thinking (Instructional Tool 2.9, p. 121)	Drawings and Annotated Drawings	♥	♥	√	√	√		♥	♥
	Concept Cartoons		♥	♥				√	
Kinesthetic Activities (Instructional Tool 2.10, p. 125)	Hands-on Experiments and Activities and Manipulatives			♥	√	♥	♥		
	Physical Movements/Gestures		♥	√	√	♥	♥	♥	

Note: A check mark (√) means that that strategy will promote student understanding, if effectively implemented. A heart symbol (♥) means that that strategy is very strongly supported by research (and is also a favorite of the authors!).

Linguistic Representations of Knowledge

The linguistic representational tools in this book examine speaking and listening (Instructional Tool 2.4), writing (Instructional Tool 2.5), and reading to learn (Instructional Tool 2.6) as ways to linguistically represent knowledge. Speaking and listening, writing, and reading are critical parts of the five essential features of inquiry (Figure 2.3, p. 27). You can use the tools during engagement and exploration, designing and conducting investigations, analyzing and interpreting data, and presenting findings and understandings (Century et al. 2002). In these tools, the authors have focused on strategies that support the development of both literacy and science literacy and that are foundational to meaning making through inquiry.

CHAPTER 2: Implementation of the Framework

Nonlinguistic Representations of Knowledge

The nonlinguistic representational tools in this book are also important for sense making. Whereas linguistic representations use language in learning, nonlinguistic representational tools store knowledge in the form of visual images. The more that students use both systems of representation, the more they will think, learn, and recall information. Instructional Tools 2.7–2.10 (pp. 107–126) will help you select appropriate strategies to promote nonlinguistic representations. Table 2.8 provides rationales for the use of the four groups of nonlinguistic representations for which we provide Instructional Tools: *models*; *maps and graphic organizers*; *drawing*; and *kinesthetic activities*.

Table 2.8

Four Selected Nonlinguistic Representations of Learning (Models, Maps and Graphic Organizers, Drawing, and Kinesthetic Activities) and the Rationales for Their Selection

Nonlinguistic Representation of Learning	Rationale	Supportive Instructional Tool in This Book
Models	We want our student to build nonlinguistic representations in their minds (Marzano 1992). Such "mental models" represent ideas, objects, events, processes, and systems. To help build and communicate mental models, people use expressed models that are simplified from those held in their brains (Hipkins et al. 2002). Types of expressed models are mathematical, physical, verbal (metaphors and analogies), visual (graphs, pictures, and diagrams), and dynamic (simulations, computer simulations, virtual manipulatives, and animations).	Instructional Tool 2.7 (p. 107)
Visual Tools: Maps and Graphic Organizers	Maps and graphic organizers are among the most commonly used strategies to help construct nonlinguistic representations. They combine linguistic and nonlinguistic modes because they call for words and phrases as well as symbols. When used as advanced organizers, they can help students retrieve what they already know, thus activating their prior knowledge (Marzano, Pickering, and Pollack 2001). Furthermore, these visual thinking tools are used for storing ideas already developed and for construction of content knowledge.	Instructional Tool 2.8 (p. 114)
Drawing	Student drawings can be used to determine students' levels of conceptual understanding, observational skills, and abilities to reason, as well as their beliefs. Teachers can use drawings as learning experiences or assessments (McNair and Stein 2001).	Instructional Tool 2.9 (p. 121)
Kinesthetic Activities	Kinesthetic activities involve movement. In such activities, there is a constant interplay between movement and learning (this interplay can even occur in our adult lives). The association of movement with specific knowledge produces a mental image of that knowledge in the learner's mind (Marzano, Pickering, and Pollack 2001). Teachers need to purposefully integrate movement into their everyday instruction.	Instructional Tool 2.10 (p. 125)

CHAPTER 2: Implementation of the Framework

The next section, where we look at Stages VII–X for implementing the Instructional Planning Framework, guides you through strategy selection for use in the responsive phase for Learning Target #1. As you select strategies, complete the Strategy Selection Template (Table 2.9) for the topic of the flow of matter and energy in ecosystems. The selected strategies for inquiry and metacognition are already filled in.

Table 2.9

Strategy Selection Template for "The Flow of Matter and Energy in Ecosystems"

	Learning Target #1	**Learning Target #2**	**Learning Target #3**	**Learning Target #4**
	Food is a substance that provides fuel and building materials for organisms.	Plants need sunlight, air, and water to make food and to grow.	Animals get food either by eating plants directly or by eating animals that ate plants.	Some organisms use waste materials and dead organisms for food, decomposing both and returning matter to the soil.
Possible Strategies for:				
Identifying Preconceptions				
Eliciting and Confronting Preconceptions				
Sense Making				
Demonstrating Understanding				
Selected Inquiry Strategy	Give priority to evidence in responding to questions (i.e., when students respond to questions from their peers or their teacher, they should first use any evidence they have gathered to support their answers).			
Selected Metacognitive Strategy	Plan and self-regulate.			

CHAPTER 2: Implementation of the Framework

Stage VII: Select Strategies to Identify Your Students' Preconceptions

You need to consider students' preconceptions because learning occurs when students make connections and construct patterns—something that depends on their prior knowledge (Lowery 1990). Although it is true that you should gear your lessons to your students' developmental levels and provide them multiple pathways to understanding (Weiss et al. 2003), you must also know their preconceptions in order to design instruction that helps them examine their misconceptions.

You have already become familiar with the misconceptions that children in general bring to the topic of the flow of matter and energy (misconceptions that are documented in the research literature; refer back to Table 2.7). It is equally important to determine *your own* students' preconceptions because misconceptions can vary by age, sex, geography, and student motivation or interest (Westcott and Cunningham 2005). The following process helps you identify your students' preconceptions well ahead of actual instructional time so you can take their thinking into account as you plan your lessons.

Step 1. Return to the Instructional Strategy Selection Tool (pp. 42–44). See column one, "Identifying Preconceptions." Scan this column on all three pages and you will find three great strategies (♥) to identify preconceptions.

Step 2. Learn more about each of these strategies by finding the appropriate Instructional Tool at the end of the chapter. Select one or two strategies that you think work well for this topic and with your students. Table 2.10, page 48, shows three strategies the authors found to be appropriate.

You might use any one of the strategies in Table 2.10 when you are planning your lessons. Annotated drawings can work especially well, but they are fairly time consuming for students to complete and for teachers to review. Anticipation guides (see Figure 2.7, p. 49, for an example), on the other hand, are quick and easy for both student and teacher. Any one of these strategies can be used to elicit preconceptions during instruction of individual learning targets, even if they were not selected for use during planning.

An anticipation guide is a quick, easy tool that activates and assesses students' prior knowledge. It is made up of carefully selected questions that can be used as a pre- and postinstruction inventory. You will notice that the statements in the sample anticipation guide (Figure 2.7) are drawn directly from the list of common misconceptions about the flow of matter and energy in ecosystems (Table 2.7, pp. 38–40). You can administer this instrument well before instruction, collect the instruments from your students, and use the results to guide your planning. (*Note:* Have students write the date in the first column when you first administer the guides. Then, when you hand them back to students after the unit, have students write in *that* date. It can be interesting to students to see how they have learned new information and changed their opinions over time.)

CHAPTER 2: Implementation of the Framework

Table 2.10

Three Strategies to Identify Students' Preconceptions

Strategy	Strengths of Strategy
Brainstorming Webs	Brainstorming webs include clustering, mind mapping, and circle maps. They allow students to generate ideas, and to brainstorm without restriction while connecting ideas. Brainstorming webs can be used to determine prior knowledge and can be revised during the course of a unit of study.
Informational Text Strategies	Some of these strategies (e.g., concept diagrams, anticipation guides, KWLs) can easily be used to identify student preconceptions. Your students may already be familiar with these tools if you use them for reading instruction. Anticipation guides and KWLs can be administered well ahead of lesson design without interrupting your current instruction, and your "findings" can be used to inform future lessons.
Drawings and Annotated Drawings	Drawing is a less-biased way than verbal literacy activities to determine understanding. It is a good strategy to see how students perceive objects, the degree to which they perceive them, and how well they can represent details. Drawings provide information about specific misconceptions, helping you address roadblocks to students' conceptual understandings. Finally, they help students grapple with their own ideas and understand abstract concepts.

For the remaining stages in the responsive phase, you identify specific strategies for each aspect of the Instructional Planning Framework (Figure 1.2, p. 8) and for each learning target. That process is modeled in this chapter for Learning Target #1 only. However, possible strategies and lesson ideas for all learning targets are found in the planning template (Table 2.15, p. 60). Before moving on, take a look at Figure 2.8, page 50, which summarizes hints and resources for Stage VII.

Figure 2.7

Anticipation Guide: "The Flow of Matter and Energy in Ecosystems"

Directions: In the first column, "Before Lessons," place a check mark (√) when you agree with the statement on the right. At the end of our unit on food chains, place a check mark (√) in the second column, "After Lessons." Then compare what your opinions about these 10 statements were before the lessons on food chains with your opinions after the lessons.

Before Lessons
(Date:_____)

After Lessons
(Date:_____)

I Agree

I Agree

_____   _____   1. Food is necessary to stay alive, healthy, and active.

_____   _____   2. Food vanishes when we use it for energy.

_____   _____   3. Animals are living things but some plants are not.

_____   _____   4. What we eat becomes part of our muscles and the rest of our bodies.

_____   _____   5. Plants use materials in the soil for food.

_____   _____   6. Animals need air but plants do not need air.

_____   _____   7. Matter in a plant is later found in animals that ate the plant.

_____   _____   8. Dead animals' bodies disappear into the soil.

_____   _____   9. Many kinds of organisms decompose dead plants, animals, and waste.

_____   _____   10. Decomposed matter in soil becomes part of plants that grow there.

CHAPTER 2: Implementation of the Framework

Figure 2.8

Responsive Phase: Hints and Resources for Planning for Stage VII

> **Hint:** Select a strategy to determine your students' preconceptions about *all* the learning targets. Use the strategy with students well ahead of instruction and then use their responses to guide unit development. Later, for each learning target, you can use strategies to elicit preconceptions related specifically to that target.

> *Resources:*
> The Instructional Tools are your primary resources. Within each tool you will find specific resources for the strategies. Study those resources to learn more about use of the strategy and to find materials that can be readily used in your classroom. For example, text and online resources for concept cartoons are listed in the Resources section as well as on page 121 in Instructional Tool 2.9.

Stage VIII: Select Strategies to Elicit and Confront Your Students' Preconceptions

It is important to focus on students' ideas throughout your lessons. You begin by *eliciting* their preconceptions, making them public so that they, their peers, and you can grapple with thinking about the concepts. You must provide opportunities for students to express their preconceptions as well as design experiences that *confront* these preconceptions. That means you must make sure that students' alternative explanations are made public and that students have experiences that make them question their existing ideas. Remember that inquiry experiences work well in this capacity. Also recall that some strategies reviewed but not selected to identify preconceptions might work here (see Table 2.10). But there are many other strategies available that you might use, either separately or in conjunction with inquiry. Use the following three steps and the Instructional Tools to select strategies to *elicit and confront student preconceptions* for Learning Target #1.

Step 1. Review Learning Target #1 (Figure 2.1, p. 23), as well as the related misconceptions and instructional ideas in Table 2.7 (p. 38). This will help you focus on the content details of the learning target; it will also help you anticipate what learning difficulties might arise regarding these concepts.

Step 2. Refer to the Instructional Strategy Selection Tool (pp. 42–44). Go to the column headed Eliciting and Confronting Preconceptions for the strategies you need. Remember to focus on the best (♥) strategies if your time is limited.

Step 3. Find the specific Instructional Tools that provide information about the specific strategies you selected (e.g., if you selected learning logs, you would go to Instructional

Tool 2.4, Writing to Learn, p. 90, to learn more about the use of learning logs). Identify strategies you think will work best. Then refer to Table 2.11 to see the strategies identified by the authors and their comments about using them.

Table 2.11

Eight Strategies to Elicit and Confront Students' Preconceptions

Strategy	Elicit Preconceptions	Both Elicit and Confront Preconceptions
1. Learning Logs	More free-form than science notebooks, learning logs are meant to be used by students to express their observations, thus eliciting preconceptions. Learning logs can engage students in thinking about a topic because they must focus on content when writing.	
2. Science Writing Heuristic (SWH)		This is a great strategy, but it is complex. You might spend some time learning about it before using it (see Instructional Tool 2.4, p. 90, for more information). If you are already familiar with SWH, its use with the topic of the flow of matter and energy in ecosystems is appropriate.
3. Small- and Large-Group Discourse		Discourse, used with a strategy such as concept cartoons to establish the purpose of the discourse, promotes learning about concepts, metacognition, and the nature of science. It can also serve as a formative assessment. It helps students build on one another's ideas to develop explanations and plan investigations (both of which align with our inquiry focus).
4. Thinking-Process Maps		Thinking-process maps promote cognitive and metacognitive learning. They help students "learn how to learn" by having them distinguish between accurate conceptions and misconceptions. In addition, the maps tend to reduce student anxiety and increase self-confidence and motivation. They also help learners see linkages among ideas and connectivity in systems (the latter an important concept when studying the flow of matter and energy in ecosystems) better than written text.

(continued)

CHAPTER 2: Implementation of the Framework

Table 2.11 *(continued)*

Strategy	Elicit Preconceptions	Both Elicit and Confront Preconceptions
5. Visual Models		Though a good strategy, visual models do not work particularly well with this content (the flow of matter and energy in ecosystems).
6. Dynamic Models	This strategy has potential with this content. However, dynamic models are better used with content that is best understood through visualization, such as how balls move on ramps.	
7. Drawing	This is a great strategy, as discussed in Table 2.10 (p. 48).	
8. Concept Cartoons		Concept cartoons work very well because they draw from different "intelligences" and are a safe way for students to express alternative explanations.

As you can see in Table 2.11, there are many effective linguistic and nonlinguistic strategies, and it is good to use a mix of them. Concept cartoons or annotated drawings used in conjunction with student discourse are especially effective. Concept cartoons present various explanations from which students can choose; students then discuss the reasoning for their choices. Not only does a cartoon present various explanations in a safe environment, the exact purpose for which you elicit and confront preconceptions, but it also provides a meaningful context for discourse. In addition, a cartoon can help students identify questions they can later conduct an experiment to answer. Remember that the inquiry focus that we selected requires that students find ways to gather evidence to respond to questions—which means they need questions to consider!

Stage IX: Select Sense-Making Strategies

Once you provide experiences (via your strategies) to elicit and confront students' preconceptions, you need to give students opportunities to make sense of those experiences. Because students benefit from multiple exposures to content, select three or four sense-making strategies. Having extra strategies on tap is also useful if your formative assessment results require that you provide additional learning experiences for students. Furthermore, a variety of strategies help you differentiate instruction, re-teach certain aspects of the topic, and/or set up learning stations (see Chapter 3 for ideas).

Step 1. You use the same approaches used in Stage VIII, but this time with a focus on sense making. That means you go to the column headed Sense Making in the Instructional Strategy Selection Tool, p. 42) for possible strategies. Notice that this column is divided into three subcolumns: Perceiving, Interpreting, and Organizing Information; Connecting Information; and Retrieving, Extending, and Applying Information (and Using Knowledge in Relevant Ways). Since we are working on the first lesson in the unit of study, you might focus on the first and second of these subcategories, knowing that later lessons will build on this lesson and apply what students learned to further learning experiences. A quick look at the strategies under Perceiving, Interpreting, and Organizing Information tells you that writing-to-learn strategies and some reading-to-learn strategies are especially effective (♥) as are some models and visual tools.

Step 2. Now refer to the Instructional Tools (pp. 76–126), which tell you more about the strategies you identified in Step 1 as effective for sense making. After reading the information there, decide on three or four strategies you think you might use in your initial lesson or for re-teaching, if necessary.

The authors have found multiple strategies among the linguistic representations in Instructional Tool 2.4 that work well. Both science notebooks and the Science Writing Heuristic (SWH) are possibilities. (SWH should be used only if you are familiar with the process or are willing to learn about it before implementation.) Science notebooks are a natural to fit in with investigations. Also, some of the reading tools (Instructional Tool 2.5), especially graphic organizers used for vocabulary development and informational text strategies, might be useful.

Several nonlinguistic representations can also be used. Task-specific organizers and thinking-process maps show quite a bit of promise and so are on our short list. Take a look at Instructional Tool 2.8 and you can see why. Note that the research section for Task-Specific Graphic Organizers says, "Their step-by-step nature provides concrete models, thereby providing scaffolds for students who might otherwise give up. They offer a global view of a process as well as an end point…. They also result in a written display of students' ideas, which students can reflect on and perhaps use to modify their thinking." In addition, given the inquiry focus of the unit, hands-on experiments will also work well.

Before you move ahead with the planning process, read the hints and strategies for Stages VIII and IX (Figure 2.9).

CHAPTER 2: Implementation of the Framework

Figure 2.9

Responsive Phase: Hints and Resources for Planning for Stages VIII and IX

Hint: There are numerous strategies for eliciting and confronting preconceptions and for sense making. Remember that your selection of strategies is influenced by the particular content you are teaching, where you are in the school year, and your focus areas for metacognition and inquiry. Try to use a large variety of strategies over the course of the year so that you address various learning styles and avoid repetition in lessons (repetition is sure to diminish student engagement).

Resources:
- *Differentiated Instructional Strategies for Science, Grades K–8* (Gregory and Hammerman 2008) provides a blueprint (including differentiation strategies) for improving science instruction while accommodating individual learning styles.
- *Tips for the Science Teacher: Research-based Strategies to Help Students Learn* (Hartman and Glasgow 2002) translates research into practical and easy-to-use classroom applications.
- *Classroom Instruction That Works: Research-based Strategies for Increasing Student Achievement* (Marzano, Pickering, and Pollock 2001) outlines instruction strategies to use to raise student achievement.
- http://education.jhu.edu/newhorizons/strategies/index.html is a website that provides resources about numerous strategies, including differentiation strategies.
- http://sydney.edu.au/science/uniserve_science/school/support/strategy.html is an online resource for various science teaching strategies.
- www.muskingum.edu/~cal/database/general provides general learning strategies as well as strategies specific to content areas, including science.

Hint: There are some resources that focus specifically on graphic organizers.

Resources:
- *A Field Guide to Using Visual Tools* (Hyerle 2000) is an excellent introduction to the use of visual tools for learning.
- The Graphic Organizer Website (*www.graphic.org*) includes a rich set of resources on many different types of graphic organizers.
- The SCORE website (*www.sdcoe.k12.ca.us/SCORE/actbank/torganiz.htm*) includes both a teacher activity bank and a student activity bank about graphic organizers.
- The Education Oasis website (*www.educationoasis.com/curriculum/graphic_organizers.htm*) provides PDFs of 58 different graphic organizers.
- *Developing Visual Literacy in Science, K–8* (Vasquez, Comer, and Troutman 2010), in addition to discussing and displaying graphic organizers, uses photographs, illustrations, diagrams, and student-created visual thinking tools to demonstrate how students can be taught to be better "readers" of graphic images.

CHAPTER 2: Implementation of the Framework

Stage X: Determine Formative Assessments

Many of the strategies found in the Instructional Tools can serve as formative assessments during a course of instruction. Your assessment findings should be used to modify instruction and support student learning throughout a unit. The steps to identify assessment strategies are the same basic steps as used in Stages VII—IX.

Step 1. Refer once again to the Instructional Strategy Selection Tool (p. 42). Look at the column headed Demonstrating Understanding. Note the two subcolumns. Strategies listed in the Formative Assessment subcolumn are intended for use directly by the teacher. Strategies in the Peer and Self-Assessment column are for students. You should include both in your lesson.

A quick scan of both subcolumns tells you that almost any of the strategies listed on the left can serve as assessments. Use a variety. Keep in mind that this unit (on the flow of matter and energy in ecosystems) is being taught early in the school year, as that might influence your selections. It also makes sense to use as formative assessments the student work that will be produced as a result of the strategies that you already built into your lesson. Why add extra assessments if the students' work will serve the purpose?

Step 2. Proceed to the specific Instructional Tools at the end of the chapter (pp. 76–126) and review the information about the strategies you have selected from the Instructional Strategy Selection Tool. Then refer to Table 2.12 to compare your choices to the potential strategies selected by the authors and the rationales for their selections.

Table 2.12

Selected Assessment Strategies and Rationales for the Selections

Assessment Strategy	Rationale for Selection
Science Notebooks	• Notebooks, already planned as one of our sense-making strategies, provide a window into students' thinking. • Extended writing allows reflection about conceptual and procedural knowledge. Thus, it is a good self-assessment. • Notebooks are most effective for assessment when the writing relates to actual science tasks that students carry out in a particular unit.
Thinking-Process Maps	Thinking-process maps can serve as evidence for the thinking of individual students or of groups; they can be revised during a lesson or unit. Perhaps you can have students generate a concept map early in this life science unit and revise the map as their thinking changes. The map could be expanded to include content related to the other learning targets as the unit progresses.

(continued)

CHAPTER 2: Implementation of the Framework

Table 2.12 *(continued)*

Assessment Strategy	Rationale for Selection
Large- and Small-Group Discourse	• Discourse is effective for both teacher assessment and for peer and self-assessment. • Discourse gives the teacher information about students' thinking, which can be used to modify instruction. • Student-to-student discourse, focused on developing explanations, allows peers to assess their own and one another's thinking.
Drawings and Annotated Drawings	• Completing the various parts of annotated drawings depends on different parts of the brain, making for rich experiences. • Drawings and annotated drawings are less biased than linguistic literacy exercises and so provide options for students who are not strong linguistic learners to show what they have learned.
Concept Cartoons	Concept cartoons can inspire discussions and written explanations that can serve as formative assessments.
Physical Movements and Gestures	A simple gesture such as a thumbs-up or thumbs-down can provide a quick sense of where students' thinking is during a lesson.

* * * * * *

That completes the process for strategy selection (see the summary in Table 2.13). Normally, you would complete the entire table, but because the process has been modeled in this chapter for Learning Target #1 only, that column alone is filled in. The selected strategies can now be used in conjunction with activities you would typically employ when teaching a unit on the flow of matter and energy in ecosystems.

Consider working with other science teachers at your grade level to generate a strategy selection template for another hard-to-teach science topic. The work of lesson planning is greatly facilitated by completion of this template. Furthermore, you will end up with valuable research-based strategies shown to be effective for helping students learn difficult science topics. The lessons themselves flow quite easily once the strategies are selected. After you go through this process once or twice, it will become second nature. Figure 2.10 (p. 58) summarizes our hints and resources for Stage X.

Table 2.13

Strategy Selection Template for "The Flow of Matter and Energy in Ecosystems"*

	Learning Target #1	**Learning Target #2**	**Learning Target #3**	**Learning Target #4**
	Food is a substance that provides fuel and building materials for organisms.	Plants need sunlight, air, and water to make food and to grow.	Animals get food either by eating plants directly or by eating animals that ate plants.	Some organisms use waste materials and dead organisms for food, decomposing both and returning matter to the soil.
Possible Strategies for:				
Identifying Preconceptions	Anticipation guide			
Eliciting and Confronting Preconceptions	Concept cartoons or probes with small- and large-group discourse			
Sense Making	Hands-on experiments, science notebooks, task-specific organizers, small- and large-group discourse			
Demonstrating Understanding	Student planning charts; science notebooks, including student explanations, thinking-process maps, annotated drawings, and concept cartoon responses; physical gestures like thumbs-up and thumbs-down; anticipation guides as post-lesson inventories			
Selected Inquiry Strategy	Give priority to evidence in responding to questions (i.e., when students respond to questions from their peers or their teacher, they should first use any evidence they have gathered to support their answers).			
Selected Metacognitive Strategy	Plan and self-regulate.			

Note: This table is the same as Table 2.9, but in this table, the strategies for Learning Target #1 are filled in.

Figure 2.10

Responsive Phase: Hints and Resources for Planning for Stage X

Hints: Formative assessments are assessments *for student learning*. They are not to be used to determine if students have the right or wrong answer but rather to determine their understanding of a concept. Formative assessments can be planned and organized or they can be quick assessments (e.g., you ask a student a question or you glance at a student's written work). You use formative assessments to modify your instruction to ensure that every student learns. Many of the strategies in the Instructional Tools can also serve as formative assessments.

Resources:

- *Science Formative Assessment: 75 Practical Strategies for Linking Assessment, Instruction, and Learning* (Keeley 2008) includes 75 specific techniques that help science teachers determine students' understanding of important ideas and design learning opportunities to meet students' needs.
- The six books in the NSTA Press *Uncovering Student Ideas* series provide a variety of (a) formative assessment probes you can use to determine your students' preconceptions of fundamental science concepts and (b) suggestions to address students' ideas and promote learning (Keeley 2011; Keeley, Eberle, and Farrin 2005; Keeley, Eberle, and Dorsey 2008; Keeley, Eberle, and Tugel 2007; Keeley and Harrington 2010; Keeley and Tugel 2009. See information on all six books at *www.nsta.org/publications/press/uncovering.aspx*).

TIME FOR REFLECTION

Take a moment and reflect on the process for the responsive phase. In Chapter 1 (Table 1.3, p. 13), you took a personal pre-assessment about your understanding of the Instructional Planning Framework components, and you used your responses to establish two learning goals. Consider your current level of understanding about the components and any questions you might still have. Respond to Table 2.14. When you work through Part II of this book, continue to focus on the aspects of the process and the tools that will help you learn more about your remaining questions.

Table 2.14

Quick Check on Your Personal Learning Goals (Refer to Table 1.3)

Personal Learning Goal	Current Level of Agreement*					Things I Have Learned	Questions I Still Have
Goal 1:	SD	D	U	A	SA		
Goal 2:	SD	D	U	A	SA		

* SD = strongly disagree, D = disagree, U = undecided, A = agree, SA = strongly agree

Resulting Lessons for Learning Targets #1–4

You have completed strategy selection for the first learning target in the instructional unit focused on the flow of matter and energy in ecosystems. Now you can apply the strategies to activities previously identified as aligned with Learning Target #1. This section describes the lesson (instructional plan) that is briefly outlined for Learning Target #1 in the Planning Template (Table 2.15, p. 60). Instructional plans for the other three learning targets are not described in the text, but they are fully outlined in Table 2.15. The same process described in this chapter for Learning Target #1 was used by the authors to develop the instructional plans for Learning Targets #2–4.

CHAPTER 2: Implementation of the Framework

Table 2.15

Responsive Phase: Planning Template for "The Flow of Matter and Energy in Ecosystems" (With Instructional Plans for Learning Targets #1–4)

Identify Student Preconceptions	Have students fill out the anticipation guide (Figure 2.7, p. 49) well ahead of your lesson planning. Use the results to specifically target your students' learning needs.

Learning Target #1:
Food is a substance that provides fuel and building materials for organisms.

Instructional Plan for Learning Target #1
(This plan does not account for changes that will come about during instruction or in re-teaching.)

PART 1

Elicit and Confront Preconceptions
Complete an index card sort with two categories—*food* and *not food*—to elicit student preconceptions and reach small-group consensus on the definition of *food*.

Provide Sense-Making Experiences
Reach class consensus on the definition of *food*. Compare the class definition to the scientific definition (i.e., *food* means materials that contain chemical energy for living things and that help them live and grow). Using Nutrition Facts from food labels, clarify that calories are a measure of energy in food.

PART 2

Elicit Preconceptions
Elicit individual student responses to a concept cartoon (Figure 2.11, p. 69).

Confront Preconceptions
1. Use small-group discussions to confront student conceptions about the cartoon (groups reach consensus on a response and clarify reasons for the choice).
2. Facilitate a whole-class discussion about these responses and generate questions that students might investigate. Guide the inquiry by identifying the question for students to investigate. This ensures that their work will address the learning target.
3. Construct a concept map that reflects the group's current thinking about how food, once eaten, impacts the body.
4. Make a group claim about the question for investigation and record it in science notebooks.
5. Provide students a limited set of materials to guide investigations.

CHAPTER 2: Implementation of the Framework

Table 2.15 *(continued)*

> 6. Complete plans for investigation. (Stress the importance of the inquiry goal for this lesson—students provide evidence when they are answering the questions of their peers and teachers—and urge students to think carefully about how they will gather the evidence.)
> 7. Use a work-plan organizer (Table 2.16, p. 71) to plan the investigation (addresses the metacognitive goal: plan and self-regulate).
> 8. Complete investigations, reminding students to self-monitor by checking progress on their work-plan organizers.
> 9. Modify concept maps as the investigation continues and students' ideas change.
>
> *Provide Sense-Making Experiences*
> 1. Develop small-group explanations based on gathered evidence and share results. (Remind students to defend their explanations using evidence from their investigations.)
> 2. Facilitate a class discussion after each group's presentation that probes thinking about both the content focus and the inquiry focus.
> 3. Display the Process Tool (Figure 2.12, p. 74) and complete the inputs and outputs as a class.
> 4. Modify concept maps based on what was learned. Post them around the room, and lead a Gallery Walk to review the various maps.
> 5. Revisit the concept cartoon and have students enter individual responses in their science notebooks with explanations for their responses.
>
> *Use Formative Assessments*
> Quickly and informally assess student understanding and immediately guide instruction by questioning or probing during small-group and whole-class discussions. Students use the thumbs-up/thumbs-down signal when appropriate. Reviewing science notebooks, including concept maps and responses to the concept cartoon, completes a more formal assessment. Additional experiences to confront conceptions and for sense making should be guided by the sophistication of students' explanations of food and its purpose in animals.

Learning Target #2:
Plants need sunlight, air, and water to make food and to grow.

Instructional Plan for Learning Target #2
(This plan does not account for changes that will come about during instruction or in re-teaching.)

In this lesson, you build on students' ideas about food as a substance that provides energy and materials to build bodies. Students need to understand that plants make their own food and that plants—like animals—need food to live and grow. Part 1 is about what plants need to make food and Part 2 is about the source of matter for the plant (and eliminating soil and water as possibilities for the source of matter).

(continued)

Table 2.15 *(continued)*

PART 1

Elicit Preconceptions

1. Show a video clip that demonstrates plant growth, making certain to include images that show plant growth from a seed into a mature plant. Many videos are available through various sources, including United Streaming and Teachers' Domain. A good clip from Teachers' Domain is at *www.teachersdomain.org/resource/tdc02.sci.life.colt.plantsgrow*. After the video, ask, "How did the seed change from a seed to a seedling and finally to a sunflower?" Have students create an annotated drawing that shows their thinking.
2. An alternative is to use the probe "Is It Food for Plants?" in Volume 2 of *Uncovering Student Ideas in Science* (Keeley, Eberle, and Tugel 2007). In the list of options on page 113 in that book, add "air" and take out "carbon dioxide," "oxygen," and "chlorophyll." This probe follows well from Learning Target #1 because it uses the word *food* and explicitly probes students' thinking about what serves as food for plants and whether students understand the biological meaning of the word *food*.

Confront Preconceptions

1. Using guided inquiry, focus on the question of what plants need to grow. Have student groups grow plants under various conditions—light and water; light and no water; no light and water. They should determine how they will measure growth. Various types of measurement equipment should be made available for students to measure height and mass.
2. If time is limited or materials are unavailable, and you have a subscription to Gizmos (an online simulation site), have students complete the Growing Plants Gizmo. Find information at *www.explorelearning.com/index.cfm?method=cResource.dspDetail&ResourceID=615*.
3. In either case, students should then develop explanations about what plants require for growth.

Provide Sense-Making Experiences

Have students compare their explanations with scientific explanations (they will need to read in various sources about how scientists explain a plant's need for light and water). Discuss the fact that light is not a material that adds to plant mass.

PART 2

Elicit and Confront Preconceptions

1. Students may still think that plants get their food from the soil. Ask students where they think the plant's increasing mass came from. If they think it is the soil, there are several ways you can help dispel this notion. During the lesson, you can sprout a sweet potato in water and occasionally draw students' attention to it. You can also set up a hydroponics station. Ask students how the plant could grow and gain mass even though there is no soil. You can also have students grow radishes from seed on damp paper towels or filter paper. Students can mass out (weigh) the seeds prior to and after growth and see that the plant increased in mass even though there was no soil.
2. If students then think the mass came from water, you can repeat the radish experiment but dry the radishes and mass out (weigh) the dried radishes. There will still be an increase in mass. Note that this activity requires a digital balance to get the accuracy in measurement required to discern the change.

Table 2.15 *(continued)*

Provide Sense-Making Experiences
1. Students can now review the classic (17th-century) van Helmont experiments. Van Helmont proved that soil was not responsible for a tree's increase in mass by doing an experiment with a willow tree. He found that the tree grew by 74.4 kg without a comparable decrease in the soil's mass. Read more at *www.saskschools.ca/~pvsd/vsfprojects/ foodforlife/foodforlife/www.simplydivinecatering.com/history_of_plants.htm*. Students can then compare those early experiments to what they have studied, and further clarify their explanations.
2. Up until now, this lesson uses the term *air* instead of *carbon dioxide* and *oxygen*. For more advanced students, you can now introduce the terms *carbon dioxide* and *oxygen*. Use of these terms, coupled with the students' study of the scientific explanations, sets the stage to introduce carbon as the source of mass. Introduce *carbon dioxide* only to let students know that carbon dioxide includes carbon and that carbon becomes part of sugar and starch in plants. Do not explore the chemistry involved.

Use Formative Assessments
Assess students' understandings by having them individually complete the process tool in Figure 2.12, page 74, for photosynthesis, based on what they have learned, and write a description of what the tool represents. Collect and review this student work to determine if further instruction is required.

Learning Target #3:
Animals get food either by eating plants directly or by eating animals that ate plants.

Instructional Plan for Learning Target #3
(This plan does not account for changes that will come about during instruction or in re-teaching.)

Students now have an understanding of food and they know that plants make their own food, using it for energy and growth. They also understand that the food stored in plants is made of substances that include carbon. They are now ready to see how this food serves as a source of matter for animals in food chains. A full understanding of what happens to food in an organism's body is held for instruction in the middle school years. At this point, the concept of food established in Learning Target #1 serves as the foundation of this lesson, which looks primarily at the transfer of matter from one organism to another in a food chain.

Elicit Preconceptions
1. Hold a brief class discussion about how a puppy turns into a full-grown dog. This discussion should elicit ideas that food is required for growth. Then ask students to compare the ways that pets get their food with the ways that organisms in nature get their food.
2. Have each student complete a process tool (Figure 2.12) for a selected animal common to your area and record this information in his or her science notebook.

(continued)

CHAPTER 2: Implementation of the Framework

Table 2.15 *(continued)*

Confront Preconceptions

1. There are numerous ways to let students explore the various organisms in a food chain and learn how matter flows through the chain. This might be the point where you have students explore owl pellets, identify the various organisms eaten by the owl, and discuss where those organisms got their food—ultimately tracing back to plants.
2. Another option would be to take your students outdoors to (at the very least) make observations in an area, cataloguing the various organisms in their science notebooks. If you have more time, have students complete a field study in a nearby habitat or even in your school yard—take soil samples, survey organisms on the surface, and identify organisms in the air over the ground surface. (An outline of work involved in such a study can be found at *www2.kpr.edu.on.ca/cdciw/biomes/protocol.htm*.) In either case, have students take careful field notes in their science notebooks using words, pictures, and maps.
3. Once students have completed this fieldwork, ask them to identify who might feed on whom among the observed organisms. Then have student groups create a poster, using a sequence of process tools (see Figure 2.12, p. 74) with an image of each particular organism inside one process tool frame to sequence the organisms and indicate the flow of energy and matter along the chain.

Provide Sense-Making Experiences

1. Conduct a brief Gallery Walk, asking students to view their classmates' posters, looking for similarities and differences among the posters.
2. Share thinking as a whole group; generate lists of similarities and discuss differences that students found on their Gallery Walk. Identify key ideas learned and remaining questions.
3. To address these remaining questions, have students read various resources, using both the internet and books and periodicals, to research food chains in their community and determine the type of feeding relationships the organisms in the food chain have. Depending on the science standards in your school/district, you can vary the level of detail in this step. Students should certainly distinguish between producers and consumers. But if your standards include additional content such as primary and secondary consumers, herbivores, carnivores, and omnivores, this is the place to apply that vocabulary. You might also have students look at various food chains in different parts of your state or across the country or around the world.

Use Formative Assessments

The process tools and science notebooks used in this lesson, coupled with small-group and whole-class discussions, serve as formative assessments.

Note: The focus of this lesson is the flow of matter because most research indicates that the study of energy flow should be reserved for middle school. However, if your school/district standards at this grade level require you to cover the flow of energy, you can introduce the concept through a study of calories in food, a review of food labels, and measurement of the caloric content of various foods. The student learning and findings can then be connected to the concept of food chains.

Table 2.15 (continued)

Learning Target #4:
Some organisms use waste materials and dead organisms for food, decomposing both and returning matter to the soil.

Instructional Plan for Learning Target #4
(This plan does not account for changes that will come about during instruction or in re-teaching.)

Students now have foundational understandings of the flow of matter in food chains from plants through animals and are ready to study decomposers and their role in the flow of matter.

Elicit Preconceptions
1. Ask students to write a response to this prompt in their learning logs: "What happens to organisms when they die? Where do they go?" OR use the probe "Rotting Apple" (Keeley, Eberle, and Dorsey 2008) if you have this book in the *Uncovering Student Ideas in Science* series. Briefly discuss some of the students' responses.
2. Engage students further in uncovering their preconceptions by giving them images to sort into a graphic organizer on whiteboards or on the computer. The images should include a variety of producers and primary consumers, secondary consumers, and tertiary consumers (or *herbivores, carnivores,* and *omnivores* if these are the terms your local standards expect students to learn) as well as decomposers, even though you have not yet introduced the idea of decomposers. The graphic organizer should include column headings for each term (except *decomposers*) and a column headed "Not Certain."
3. Have each student group share their results with one other group, including the reasons for the sorts. Have them confirm the organisms that they are sure are producers and consumers and move the other organisms into the "Not Certain" category.
4. Assign partners one organism from the "Not Certain" category to research. After research, they share with the class what they learned. Students will very likely come across the term *decomposer* and can apply it to some of these organisms.
5. Tell the students that they will now learn a bit about how decomposers work.

Confront Preconceptions
1. Introduce students to stations around the classroom that have been set up to study various examples of decomposition. A variety is important because if students observe only one organism (e.g., worms) they may think all decomposers are worms (Keeley, Eberle, and Dorsey 2008). Be certain to include at least some of the following stations and/or other stations with similar examples. Try to include both plant and animal examples and at least one station where the mass of the decaying item can be measured.
    a. Bread just starting to mold (in a self-sealing plastic bag)
    b. Bananas sprinkled with yeast and misted with water (in a self-sealing plastic bag)
    c. A dead fly (in a self-sealing plastic bag)
    d. Chunks of carrots that are starting to mold in a bag with holes in it for aeration (You should start this process well ahead of this lesson by showing an initial carrot to the students, massing it out, and recording the mass. Students can then mass out the carrot, leaving the carrot in the bag, each time they observe it.)

(continued)

CHAPTER 2: Implementation of the Framework

Table 2.15 *(continued)*

    e. Worm bins with waste materials to decompose (Include objects that can be removed and measured.)

    f. Soil samples to explore, using the naked eye, hand lenses, and microscopes (Students determine the array of organisms that live in the soil and do research to determine the organisms' roles in the ecosystem.)

2. Have students rotate among stations during the course of at least a week, making thorough observations and recording both qualitative and quantitative observations in their science notebooks. Encourage them to include observations of appearance, shape, size, dimensions, and mass. Make available various measuring devices that include digital balances so that students can more accurately measure mass. Depending on the number of stations you set up and the time you have available, it may take students more than one day to visit all the stations. They should visit each station at least three times during the lesson.

Provide Sense-Making Experiences

1. Ask student groups to use the evidence they gathered to develop explanations for what happened to the materials in the various stations. Specifically, they should
   - describe and draw the physical changes in the objects;
   - identify the agent of change and why they consider it the agent of change; and
   - tell where they think the matter went, if they saw the solid material diminish.
2. Have students present their explanations to the whole class and provide the evidence for those explanations.
3. Probe their thinking to ensure that the students are building the following understandings:
   - Living organisms are all agents of change in the decay process.
   - Many different types of organisms cause these changes and some are microscopic.
   - All living things need food, including decomposers (whose food simply happens to be dead organisms and/or waste materials from living organisms).
4. Provide students with resources about what happens to carbon-based materials during decay. The intent is to help them understand that carbon-based bodies of organisms are broken down and eventually return carbon to the environment. Have them read these resources and modify their explanations (given in steps 1 and 2) based on what they have now learned. If time allows, have students apply what they learned by conducting experiments on rates of decomposition using decomposition columns (find information at *www.bottlebiology.org/investigations/decomp_main.html*).

Use Formative Assessments

Once again, many of the embedded activities serve as formative assessments. For a more formal assessment, you might pose one more scenario of decay and ask students to provide an explanation of the process along with an annotated drawing. Or, you can return to the "Rotting Apple" probe if you used it at the beginning of this lesson (for Learning Target #4).

CHAPTER 2: Implementation of the Framework

Identifying Your Students' Preconceptions About Content
Common misconceptions that children have about the flow of matter and energy were identified during the predictive phase, but you also need to know what your students' ideas are about the *content* addressed in Learning Target #1. Before planning the lesson, use the anticipation guide (Figure 2.7, p. 49) to determine your students' preconceptions. Let's say that student responses show that they do not think that food just vanishes (item #3). That is good. However, maybe they seem to have no clear understanding of how food is used to build the body and provide energy (items #1 and #3). That tells you that you must design the lesson to make sure that students' conceptions of "food" eventually come closer to the scientific explanation and that they have experiences that demonstrate that food helps an organism live and grow.

Refer to the Planning Template (Table 2.15) and notice that—for Learning Target #1—there are actually two parts in the process of eliciting preconceptions and confronting preconceptions. In Part 1, you make sure that students have a scientific understanding of the word *food* before you demonstrate how food is used. (Of course, if your students already have a solid understanding of *food* as it is used in biology, you needn't complete this part of the lesson.) In Part 2, you want to determine if students understand the role that food plays in an organism.

Eliciting and Confronting Ideas About Food (Part 1 of the Instructional Plan for Learning Target #1)
Prepare sets of index cards, one set for each student group, some cards with the names of various foods and some cards with the names of nutrients that are not food (e.g., water, minerals). Give students the following instructions:

- One student selects the first card and explains why he or she thinks it is a food or not (this activity elicits that student's preconception).
- The other students in the group either agree or disagree (their varied explanations confront the original student's preconception) and explain their reasoning.
- Group members reach consensus before a card is placed in the "food" or the "not food" pile.
- Students take turns drawing cards, explaining their thinking, and reaching consensus on card placement, until all cards are sorted.
- Once all cards are sorted, the students in each group develop a group definition of *food*.

As students complete this activity, move from group to group and listen carefully to student definitions as they evolve, but let the students have these dialogues without your input. Make note of which groups have the least and which have the most sophisticated definitions. Also note the array of definitions in between.

CHAPTER 2: Implementation of the Framework

Once all student groups have written their definitions of *food*, facilitate a whole-class discussion during which groups share their ideas. Ask the group with the least sophisticated definition to share first. Recognize (aloud) the aspects of the definition that are accurate, and ask questions that probe their less-developed conceptions, such as, "What happens to food once you eat it?" Then have the remaining groups share their definitions, following the same pattern and working toward the most sophisticated definition. Develop a class definition of *food*. Record this definition in a place where all students can readily refer to it.

Once the class reaches consensus on this definition, provide various text resources that include a scientific definition of *food* (which is generally something like "materials that contain chemical energy for living things and help them live and grow"). Have students read these definitions in their small groups and compare the class definition with the scientific definitions. Discuss the results of their comparisons as a whole class. One aspect of the definition they might miss is the concept of energy. If so, display a sample set of Nutrition Facts from a food label and explain that scientists use "calories" to measure the amount of chemical energy in a food. Also indicate that the proteins, carbohydrates, and fats listed on the labels each contain calories. At the elementary level, it would be inappropriate to explore more fully what is meant by *calories*; understanding that a calorie is a measure of energy in the food is adequate.

Now that students have a sense of what scientists mean by *food*, you are ready to provide experiences that help students determine what happens to food they eat.

Eliciting Preconceptions About the Role Food Plays in an Organism

Concept cartoons were identified as a way to elicit preconceptions (refer to Table 2.13, p. 57). A cartoon like the one in Figure 2.11 on the next page gives students optional explanations about what happens to food in an organism (the focus of Learning Target #1) rather than requiring them to develop their own. Notice that the various explanations in the cartoon reflect the common student misconceptions that had been identified in the research literature (Table 2.7).

Display this concept cartoon and have students study it. Then ask students to select which they think is the best response, record that response in their science notebooks, and give their reasoning for the selected response.

Confronting Preconceptions About the Role That Food Plays in an Organism

Organize students into three- to four-member groups and ask them to discuss their ideas about the cartoon, eventually reaching consensus on which of the cartoon choices they think is most accurate. Students' preconceptions are confronted during this conversation because not all students will make the same choice or provide the same reasoning for their choices. Their explanations may not yet align with the scientific explanation, but the cartoon helps focus their discussion on the concept of food.

CHAPTER 2: Implementation of the Framework

Figure 2.11

Concept Cartoon to Elicit Preconceptions on the Question "What Do You Think Happens to the Food That Animals Eat?"

Circulate among groups and probe student thinking, asking questions where appropriate. Your questions should target the actual misconception with which the group is grappling. Questions might include the following:

- What do you think happens to food in the stomach?
- What might happen to the food as it passes through the body?
- How does the food change to become a part of the body?
- How do you think food helps the organism move?

Once groups complete their conversations and reach consensus on an explanation, have them share their thinking with the entire class. This further confronts preconceptions as each small group shares their choices and their thinking.

Each explanation in the cartoon is also an explanation that can be tested through investigation. Have students brainstorm possible questions that might help determine which explanation is correct and record these questions for review (e.g., on a marker

CHAPTER 2: Implementation of the Framework

board, a Smart Board, or poster paper). During this class discussion, probe student thinking so that one of the brainstormed questions is something like "How does the food an animal eats affect its body?" Recall that the decision was made to use more guided inquiry (as opposed to more student-directed inquiry) in this instructional unit, so in this case you identify the question to investigate. In other words, you allow students to generate the list of possible questions, but you choose the specific question for investigation so that their work focuses on the content of Learning Target #1.

The students now plan their investigations based on the question you have chosen. Ask each group to construct a concept map of their current thinking and post it in their work area. This initial concept map serves as a record of their conceptions and is available for their review and modification as they move forward with the investigation. It also serves as an assessment for you and demonstrates their preconceptions so that you know how best to support their learning during and after the investigation. Both the students and you can periodically refer to the concept map to determine how their thinking changes.

During the investigation itself, you want students to determine how food contributes to an organism's growth. Suggest to students that they can feed an organism in the classroom over a period of time and see what happens in terms of the size of the organism. This suggestion limits the type of data students collect, moving the inquiry more to the center of the variations in inquiry (Figure 2.3, p. 27). Before students plan for an investigation of this sort, have small groups make claims in response to the question you identified ("How does the food an organism eats affect its body?"), recording them in their science notebooks. Guide them to record their claims in a form such as: "I think when we feed the organism it will ____ because ____."

You can further guide the inquiry by limiting the tools and materials available for use, helping students focus more clearly on the targeted content you are teaching. Include in the materials at least one type of organism to feed but preferably several (mealworms, crickets, or other easily maintained and fed organisms), various materials that could be used to feed and shelter the organisms, and various measuring devices.* Also include reading materials about the specific organisms that will help students in their design of the investigation. Include digital balances, if available. If you do not have them in your school, perhaps you can borrow some from your neighborhood middle or high school. Digital balances are accurate enough to measure a change in mass of a growing organism if you provide one to two weeks for the investigation.

Recall the metacognitive goal for this unit: plan and self-regulate. To address this goal, provide student groups with a copy of a work-plan organizer that includes an example related to this specific content (Table 2.16). This organizer promotes the development of planning and self-regulation skills. Ask students to complete the tasks and steps that will be required of them during their investigation. Make certain that they include preinvestigation, investigation, and postinvestigation tasks. Also recall that the inquiry goal for this unit is to give priority to evidence in responding to questions

---

*Always consider safety and hygiene issues when bringing live organisms into the classroom. At a minimum, from the point of view of your students, make sure you have a ready supply of liquid soap, paper towels, and hot water for hand washing. Also, clean desks and counters with soap and disinfectants following life science activities. For more information on the safe care and feeding of organisms in the elementary classroom see Kwan, T., and J. Texley. 2002. *Exploring safely: a guide for elementary teachers.* Arlington, VA: NSTA Press (especially pages 37–48).

CHAPTER 2: Implementation of the Framework

Table 2.16

Work-Plan Organizer for Learning Target #1 Investigation ("How Does the Food an Organism Eats Affect Its Body?")

Instructions: As you plan for your investigation, fill in the following organizer. Define the tasks you must complete in order to get ready for work. Then determine the materials needed for each task, the procedures you will follow, the team member(s) who will be responsible for doing the task(s), and the deadline to complete the task.

What Is Our Task?	What Materials Do We Need?	What Are the Procedures?	Who Will Do the Work?	By When Will the Work Be Done?
Example: Determine the organism we will use.	Reading materials about the various organisms	1. Each person reads from one source. 2. We share what we find in our reading. 3. We decide which organism to study.	The entire group	Day 1

Hard-to-Teach Science Concepts

CHAPTER 2: Implementation of the Framework

(i.e., when students respond to questions from their peers or their teacher, they should first use any evidence they have gathered to support their answers).

Even though you identified the question for the investigation, students determine what serves as appropriate evidence. Their work plans should include materials and steps that lead to appropriate evidence. For instance, if they want to know what happens to the organism's size (e.g., length, mass) during the investigation but do not include appropriate measuring devices (in the "What materials do we need?" list) to gather those data, there is a problem. As students complete their plans, circulate around the room and ask students questions (e.g., "How will you know that the organism grew?") to ensure they are thinking about the need for measuring devices.

Give students adequate time each day for at least one week (preferably two) to observe the organisms they investigate, and provide time two to three times during the week for them to take measurements. Remind students to carefully record all data.

Student experiences during the investigation will continue to confront their preconceptions. The results might surprise them or, at least, require that they refine their explanations. This aspect of investigation aligns with the second and third essential features of inquiry: formulating explanations from evidence and using that evidence in responding to questions from their peers and the teacher (as noted in Figure 2.3, p. 27). During the investigations, they should continue to develop their own explanations. Only later in the lesson will you help them connect their explanations to scientific knowledge (the fourth essential feature of inquiry).

Sense Making About the Role of Food in an Organism's Body

Sense making actually begins when the experimental evidence either confirms the students' original conceptions or challenges their thinking. Students use the data to develop and refine explanations, and you facilitate this process through your questioning of the group as they work. Remind students to record their evolving explanations in their science notebooks. Circulate around the classroom, determine the degree of understanding in each group, and ask probing questions such as the following:

- What evidence do you have that the animal grew?
- How do you know that this occurred because of what you fed them?

Also, remind students to modify their concept maps if their explanations change. You can quickly scan these maps and determine how they reflect students' thinking and how they align with the explanations they provide as you ask questions. Pay special attention as you move among groups to determine those students with the most naive explanations and those with the most sophisticated explanations. You will use this information later when you have the groups share their results.

One major way in which scientists make sense of their investigations is to formulate explanations and share them with other scientists. In the same way, students should share their explanations with other students. Ask students to prepare their explanations to share with the group and remind them that they should be able to defend their explanations based on the evidence they gathered. You might scaffold this process by asking each group to prepare a statement based on this frame: "I used to think _____ because _____, but now I know _____ because _____."

Ask the group with the most naive explanation to present first, working toward the presentation of the group with the most sophisticated understanding. Focus on student conceptions, providing feedback that reinforces accurate thinking and asking questions about lingering misconceptions (e.g., "So you think the food provided material for the animal to grow but you aren't certain. What makes you think that? What evidence do you have that this might be true?"). Then proceed with each group, making certain to probe students' thinking about why the organisms' mass increased. Build toward the explanation that best aligns with that of scientists. This allows every group to contribute to the growing understanding developed during the presentations.

As students share experimental results, your facilitation should also focus on how well students (a) understand which data collection methods align best with the question they are attempting to answer and (b) generate data most effective at building an explanation (the inquiry focus). For example, ask questions of the group such as, "Why did you measure length instead of mass?" or "Which of your data best support that answer?"

Now use the process tool shown in Figure 2.12, page 74. In the center, add an image of a baby and of a child that the baby eventually became. These images represent human growth. Display the tool and discuss with the class the matter and energy inputs and outputs. At this point, students should understand that food provides the matter and energy inputs, though there has been no discussion of air and its importance to animals. Students generally know that the calories in food are measures of the available energy and that liquid or solid food provides matter. This matter is then used and increases the mass of the organism (more cells as output causing growth as a matter output). At this point in the lesson, forms of energy and energy outputs have not been discussed, but you can probe student thinking about these ideas. (Also, students have not yet studied what actually happens to food during digestion and assimilation.)

Demonstrating Understanding of Concepts

Ask students to revisit their concept maps one more time and modify them as necessary, drawing them on large poster paper to display. Conduct a Gallery Walk, moving from poster to poster with the class. Discuss each poster with the students, clarifying those concepts that might need elucidation. Finally, show the concept cartoon once again and ask students to record their current responses to the cartoon and their reasoning in their science notebooks.

CHAPTER 2: Implementation of the Framework

Figure 2.12

Matter and Energy Process Tool

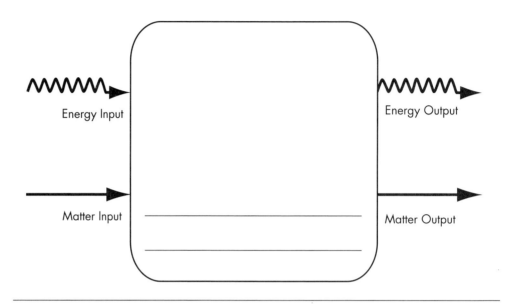

Source: Mohan, L., and C.W. Anderson. 2009. Teaching experiments and the carbon cycle learning progression. Paper presented at the Learning Progressions in Science (LeaPS) conference in Iowa City, IA. Reprinted with the permission of the authors.

It is important to note that this lesson, based on research-based strategies, is but one possible lesson on the flow of matter and energy. There are many ways to approach this content, just as there are many effective strategies. It is also possible that your students will not fully understand the concepts even after instruction, and you will have to consider other approaches to re-teach. It is important to select multiple strategies not only for use during initial (planned) instruction, but also for possible follow-up teaching experiences, as determined by formative assessments. Re-teaching and follow-up teaching are examples of the iterative nature of the responsive phase, further discussed in Chapter 3.

Endnotes

[1]Current research on learning was used to develop Part I of this book, specifically in the creation of the 10 Instructional Tools. Students' conceptual learning is informed by research related to three perspectives: (1) cognitive perspectives that focus on individual conceptual change, (2) sociocultural and social constructivist perspectives that recognize the importance of conceptual change but view the social context as central to the learning process, and

CHAPTER 2: Implementation of the Framework

(3) participative perspectives that view learning as a process best established through an apprenticeship model. For a more thorough discussion of each approach, their strengths and weaknesses, and ways in which each contribute to our thinking about learning science, see the essay "Student Conceptions and Conceptual Learning in Science" (Scott, Asoko, and Leach 2007). Curriculum and instruction are likely informed to some degree by each of these perspectives (Carlsen 2007; Scott, Asoko, and Leach 2007). Our agreement with this contention is evident in the strategies included in the Instructional Tools.

[2] Conceptual change in our students depends on their acquiring new knowledge (cognition) as well as on their reflections on their learning (metacognition). Such reflections lead students to detect and monitor incongruities in their own understanding, alerting them to their learning difficulties. Because elementary students have greater metacognitive abilities than previously thought and they can certainly become more metacognitive, we need to tap into those abilities so they can help guide their own learning (Duschl, Schweingruber, and Shouse 2007). Inquiry and the nature of science are also central to conceptual understanding. We cannot simply tell students what they need to know. They must make sense of the content themselves, and inquiry is central to meaning making in science. We will discuss metacognition and inquiry further when we introduce the tools available for use in the responsive phase.

CHAPTER 2: Implementation of the Framework

Instructional Tool 2.1

Teaching the Five Essential Features of Inquiry

1. Engage in Scientifically Oriented Questions

What Does This Mean and Why Is It Important?

Questions in science look for patterns in nature in order to develop explanations for observed phenomena. They help students focus on core concepts (NRC 2000). When students generate their own questions, this generation probes their understandings; improves content retention, conceptual understanding, and problem-solving abilities; and guides inquiries (Colbert, Olson, and Clough 2007; Wright and Bilica 2007).

Classroom Implications

- Have students interact with phenomena in order to establish a question focus rather than an answer focus. As students interact with materials, questions arise that are derived from their prior knowledge and the interaction itself. Use starters such as laboratory explorations, demonstrations, reading a scientific study, and community observations to generate a range of questions.
- Students may be used to highly structured activities, so they will need support as they begin their investigations (Hipkins et al. 2002). You can improve questioning by encouraging students to formulate their own questions and responding positively to students' spontaneous questions. You can provide examples of testable questions, materials that stimulate questions, opportunities to explore information related to their questions, and feedback and the opportunity to change factual questions into testable questions or to generate new questions (Chin, Brown, and Bertram 2002; Harlen 2001; Krajcik et al. 1998).
- The Visible Thinking website includes a routine called Think/Puzzle/Explore, which activates prior knowledge, generates ideas and curiosity, and sets the stage for inquiry. The routine can be used with the whole class to generate possible ideas for inquiry. Students write down their own ideas and then share them with the class; you write down the ideas on a class list that everyone can see. (See *www.pz.harvard.edu/vt* for Think/Puzzle/Explore).
- Posed questions are not all investigable. Some are too broad to investigate and some are too narrow and can be quickly answered by using reference materials. "Why" questions are common but often not investigable; "how" questions, on the other hand, lend themselves to inquiry (NRC 2000). A questioning tree that helps students generate engaging and investigable questions is found at the Southwest Center for Education and the Natural Environment (SCENE) website (*http://scene.asu.edu/habitat/inquiry.html#4*). Questions are sorted first by unanswerable (why) or answerable (how, what, when, who, which), then by interesting or uninteresting, next by comparative or noncomparative, and finally by manipulative or observational (Southwest Center for Education and the Natural Environment 2004).

Instructional Tool 2.1 *(continued)*

Application Example
Have students work with materials related to the targeted content (e.g., temperature and states of matter). Set up stations at which students interact with materials designed to explore various aspects of the content, make observations, and generate questions. Work with students to identify which questions are investigable, which are not, and which might be changed to make them investigable (e.g., they can change "why" questions into "how" questions or they can narrow broad questions, making them easier to control).

Technology Applications
• Inspiration software can be used to brainstorm/generate questions related to content investigations. In addition, questions that are informative and perhaps foundational to the content, but not investigable, might be answered via a web search. This can be done individually, in small groups, or as a whole class. • E-mail/webconferencing/blogging with scientists and other students can be used to generate questions. • Microblogging, like Twitter (*http://twitter.com*), can be used to capture students' questions.

2. Give Priority to Evidence in Responding to Questions

What Does This Mean and Why Is It Important?
Science uses empirical evidence as the basis for explanations about how the world works. Student work should reflect this. Evidence includes data and measurements from observation of phenomena. Human senses and specialized instruments are used to collect data. Sometimes conditions can be controlled and other times they cannot. Data are verified for accuracy by checking measurements, repeating observations, and gathering various data around the same phenomenon. Evidence can be further questioned and investigated. Students can engage in all of these actions in order to learn to use evidence when responding to questions from their peers or the teacher (NRC 2000).

Classroom Implications
What is meant by *evidence*? Scientists concentrate on getting accurate data from observations of phenomena. They obtain evidence from observations and measurements taken in natural settings or in settings such as laboratories. They use their senses, instruments to enhance their senses, or instruments that measure characteristics humans cannot sense. Sometimes scientists can control conditions to obtain evidence. Sometimes, they cannot, so they gather data over a range of naturally occurring conditions and over a period of time in order to infer what the influence of different factors might be. Accuracy is verified by checking measurements, repeating observations, or gathering different kinds of data related to the same phenomenon. The evidence is subject to questioning and further investigation.

(continued)

CHAPTER 2: Implementation of the Framework

Instructional Tool 2.1 *(continued)*

You want students to use evidence to develop explanations when they conduct classroom inquiries. Specifically, you may want them to

- observe and carefully describe characteristics of plants, animals, and rocks.
- take measurements of temperature, distances, and time and carefully record these measurements.
- observe matter transformations and Moon phases and chart progress.
- obtain evidence from the teacher, instructional materials, or the web.

Application Example

Students design an experiment around a particular hypothesis. They begin the design process by diagramming what evidence would be necessary to confirm their prediction, what evidence would refute it, and what types and amounts of data need to be generated to provide that evidence. They gather the appropriate data and use the data as evidence to formulate their explanations.

Technology Applications

- Free graphing software is available, including at *http://nces.ed.gov/nceskids/ createagraph*. The Computational Science Education Reference Desk (*www.shodor.org/ refdesk*) is a wonderful resource for free tools for creating graphs, calculators that plot changes between dependent variables, lots of simulations and computational models, and software for creating computational models.
- InspireData (Inspiration Software, Inc.) lets students explore and analyze data with a variety of tools. Databases are available, but you can also create databases using the InspireData e-Survey tool or import internet databases. Information is found at *www. inspiration.com/productinfo/inspiredata/index.cfm*.
- Teaching Science Through Inquiry With Archived Data (Haury 2001) presents two strategies for engaging with data via the web to extend inquiry; it also includes links to sites with archived data. This ERIC Educational Report can be found at *http://findarticles. com/p/articles/mi_pric/is_200112/ai_1912240294*.

Instructional Tool 2.1 *(continued)*

3. Formulate Explanations From Evidence

What Does This Mean and Why Is It Important?

This aspect of inquiry differs from #2 in that it emphasizes the path from evidence to explanation rather than the criteria for and characteristics of the evidence (NRC 2000). Explanation consists of *making a claim, supporting the claim with appropriate and sufficient evidence*, and *providing reasoning that links the claim with evidence and explains why the data are evidence in support of the claim* (McNeill and Krajcik 2008). Cognitive processes used in formulating explanations include classification, analysis, inference, and prediction. General processes such as critical reasoning and logic are also used. Explanations require students to build new ideas on current understandings. Content learning increases when students think about whether evidence does or does not support their personal theories (Hipkins et al. 2002; NRC 2000).

Classroom Implications

- You can't assume that your students know how to write explanations, so you should provide explicit instruction. Students struggle more with supporting a claim with evidence and with reasoning than they do with making a claim. You should provide specific definitions and examples of what is meant by *claim* and *evidence* and let students discuss what they think *evidence* means. You should model and critique explanations, either in written or verbal form. Five general teaching strategies that help in student formulation of explanations are (1) making the framework—claim, evidence, and reasoning—explicit; (2) modeling and critiquing explanations; (3) providing a rationale for creating explanations; (4) connecting scientific explanations to everyday explanations (e.g., "How might you convince your parent that you need a higher allowance?"); and (5) assessing and providing feedback to students (McNeill and Krajcik 2008).
- Even if you do not conduct full inquiries, your students can improve their abilities to formulate explanations from evidence if you provide them with raw data and primary sources from which to develop explanations. Reading, discussion, and research are also beneficial (Krueger and Sutton 2001). Concept cartoons can sharpen the conceptual and metacognitive focuses during argumentation (Keogh and Naylor 1999, 2007). (See pp. 123–124 for further information on concept cartoons.)

Application Example

The Visible Thinking website includes a thinking routine called What Makes You Say That? It is a flexible thinking routine that asks students to describe something and then to support their interpretations with evidence. It can be used to make scientific observations and hypotheses (see thinking routines, specifically What Makes You Say That?, at *www.pz.harvard.edu/vt*).

(continued)

CHAPTER 2: Implementation of the Framework

Instructional Tool 2.1 *(continued)*

Technology Applications

- InspireData (Inspiration Software, Inc.) lets students explore and analyze data using various tools. Databases are available in the software, but you can also create databases using the InspireData e-Survey tool or import internet databases. Information is found at *www.inspiration.com/productinfo/inspiredata/index.cfm*.
- Haury (2001) suggests two strategies to engage with data via the web: access data sets built by science projects or agencies and collaborate with other school groups to generate data. He provides a list of available resources as well (*http://findarticles.com/p/articles/mi_pric/is_200112/ai_1912240294*).
- Digital libraries are good tools for finding scientific data sets. One example is the Science Education Resource Center's Using Data in the Classroom portal (*http://serc.carleton.edu/usingdata/index.html*), which has worthwhile resources such as data-set-integrated lessons.
- Students can put georeferenced data that they have generated into Google Earth, a good data visualization exercise.
- Wikis can be used to record and analyze data.

4. Connect Explanations to Scientific Knowledge

What Does This Mean and Why Is It Important?

As noted in #3, explanation includes three components: making a claim, supporting the claim with appropriate and sufficient evidence, and providing reasoning that links the claim with the evidence and explains why the data serve as evidence to support the claim (McNeill and Krajcik 2008). The third of these components is what is done to connect explanations to scientific knowledge (principles).

Classroom Implications

- Students can review alternative explanations when they engage in discussion, compare results, or check their results with explanations provided by you or instructional materials (NRC 2000). It is important to make sure students connect their results with currently accepted scientific knowledge appropriate to their developmental levels.
- You should provide specific definitions and examples of what is meant by *reasoning*. It is important to hold specific discussions that help students understand that they need to write down the underlying scientific principles they use to select their evidence (McNeill and Krajcik 2008). It is also important that you provide well-structured materials that make evidence for scientific theories apparent so that students have examples of what assertions based on evidence look like (Hipkins et al. 2002).

Instructional Tool 2.1 *(continued)*

Application Example
Students in a group have just completed an inquiry using cars and ramps and have developed their best explanation based on the evidence they gathered. Their teacher now uses a jigsaw activity—she distributes a *different* text resource with a scientific explanation to *each* member of the group. Each group member reads his or her assigned text and writes a summary of the similarities and differences between the explanation in the text and the group's earlier explanation. Then each group member shares his or her findings with the group. The group then talks about possible modifications to their explanation and generates new questions. If necessary, they return to the car and ramp activity and explore these questions.
Technology Applications
• Once again, digital libraries are rich resources. • Both online research and electronic communication with science experts support this feature of inquiry. Many organizations support this type of work, including the Globe Program, which is a premier example and actually supports each of the five features of inquiry. Information can be found at *www.globe.gov*. • Newton's Ask a Scientist website is a good resource for students (*www.newton.dep.anl.gov/aasquesv.htm*).

5. Communicate and Justify Explanations
What Does This Mean and Why Is It Important?
Students should clearly communicate their explanations in such a way that the work can be replicated. This requires communication of the question researched, procedures followed, evidence gathered, the explanation proposed, and review of possible alternative explanations. These explanations are open to review. Sharing explanations helps students question and/or reinforce their understandings about evidence, scientific knowledge, and their explanations.
Classroom Implications
Have students share explanations. Peers should "ask questions, examine evidence, identify faulty reasoning, point out statements that go beyond the evidence, and suggest alternative explanations for the same observations" (NRC 2000, p. 27). This process helps them make connections among evidence, scientific knowledge, and their explanations. An authentic argument is made up of (1) a potential explanation, (2) the data that support the explanation, (3) a summary of other possible explanations, and (4) if required, an explanation of how the initial model changed in light of the evidence (Windschitl 2008).

(continued)

Instructional Tool 2.1 *(continued)*

Application Example
Student groups give poster presentations that summarize long-term environmental research in a habitat near their school. They first give presentations to their fellow students, communicating and justifying their explanations. They are critiqued by their peers and then given time to finalize their posters and, as necessary, reconsider their explanations. They then present their results at a community forum that includes scientists, parents, and community members. They once again present their findings, sharing alternative explanations and justifying their explanations.
Technology Applications
Numerous applications, including blogs and presentation software, can be used to communicate and justify explanations.

Instructional Tool 2.2

Three Strategies That Support Metacognition

1. Critical Thinking and Learning	
Visible Thinking: Truth Routines Truth Routines help students identify truth claims and explore strategies to uncover the truth, think more deeply about the truth of something, clarify claims and sources, explore truth claims from various perspectives, and determine the various factors relevant to a question of truth and see beyond an either/or approach to truth. These routines promote critical thinking because they encourage students to seek accuracy and clarity, be open-minded, and restrain impulsivity. Explore the Visible Thinking website (*www.pz.harvard.edu/vt*) to learn more.	
The Research	The Visible Thinking website is based on years of research about thinking and learning as well as research and development in classrooms. All strategies shared on the site were developed in classrooms and were revised multiple times to ensure that they were applicable in the classroom and that they promoted student thinking and engagement. These strategies allow students with disabilities and general education students to show what they know, provide gifted and talented students a way to explore ideas at a deeper level, and give teachers a window into their students' thinking, allowing them to provide targeted and differentiated instruction. Visible thinking changes the nature of classroom discussions, making them more student-directed (i.e., increasing students' active participation) and inclusive (i.e., making sure everyone feels safe about participating) (Ritchart and Perkins n.d.).
Classroom Implications	Some of these strategies work very well during inquiry when you want your students to formulate explanations from evidence and connect their explanations to scientific knowledge. They also work well when exploring controversial issues related to science. You can use the various strategies with individuals and small groups and during whole-class discussion.
Application Example	One strategy at the website is Claim/Support/Question. The student makes a claim, identifies support for the claim, and then questions the claim. You can use this strategy during initial explorations of specific content. For example, set up a series of stations with stream tables to explore erosion—one station might look at slope and another at rate of water flow. After visiting the stations, each student makes a claim related to his or her observations and identifies support for the claim that is based on things that can be seen, felt, and known. Then other students ask questions about the original student's claim—for example, about what isn't explained or what's left hanging. These questions point students to areas of experimentation.

(continued)

CHAPTER 2: Implementation of the Framework

Instructional Tool 2.2 *(continued)*

Visible Thinking: Fairness Routines	
Fairness Routines help students explore diverse perspectives, consider attitudes and judgments, separate fact and feeling, and explore the complexity of dilemmas. They promote critical thinking because they relate to open-mindedness and sensitivity to others' feelings. Explore the Visible Thinking website (*www.pz.harvard.edu/vt*) to learn more.	
The Research	See Fairness Routines on the Visible Thinking website.
Classroom Implications	These routines are easy to use and can be infused into all kinds of content instruction. The Circle of Viewpoints strategy is useful to begin discussions about controversial issues and dilemmas. Reporter's Notebook helps students separate fact from feeling and might be useful when teaching science concepts that conflict with students' beliefs and feelings.
Application Example	The Circle of Viewpoints strategy can be used in an ecology unit that looks at water pollution. Students brainstorm a list of perspectives related to the issue. Then each student selects a viewpoint, describes the topic from that viewpoint, and asks a question from that viewpoint. Finally, students consider new ideas and questions they now have about the topic as a result of having done the activity.
Resources	See graphic organizers in Instructional Tool 2.8.

2. Self-Regulated Thinking

Identify What You Know and What You Don't Know*	
At the beginning of an activity, students identify both what they know and what they don't know about a topic. As they research and learn about the topic, they verify, clarify, expand, or replace their original thinking (Blakey and Spence 1990).	
The Research	• Personal goals are important because we all need to plan and manage resources (Marzano 1992). • Concept mapping has been shown to benefit students' metacognitive abilities (Stow 1997). • Concept cartoons increase students' awareness of their ideas (Keogh and Naylor 1999).
Classroom Implications	• Concept mapping coupled with interviews help students analyze their thinking, identify their strengths and weaknesses, and set learning targets (Stow 1997). • 3-2-1 Bridge and Generate, Sort, Connect, Elaborate are visible thinking routines that help activate prior knowledge and make connections (Ritchart and Perkins 2008) (Visible Thinking website: *www.pz.harvard.edu/vt*). • Many graphic organizers require students to identify what they know and what questions they have about a topic. (KWL is probably the most common example.)

*All strategies in this section—2. Self-Regulated Thinking (pp. 84–88)—followed by an asterisk (*) are outlined in Blakey, E., and S. Spence. 1990. (ERIC Digest no. ED327218) *www.ericdigests.org/pre-9218/developing.htm*.

Instructional Tool 2.2 *(continued)*

Application Example	Use a probing question to start a lesson on just about any concept. Have students respond to that question using either a mapping tool or a KWL chart. Students then reflect on their individual charts or maps and identify personal learning goals for the lesson.
Technology Applications	You can use the computer program Inspiration to generate initial ideas and questions. Then flag and annotate areas of digital documents to mark questions and capture unfamiliar concepts and terms. Helpful tools include digital sticky notes (Google "sticky notes" for multiple links), highlighting and commenting tools in Word or Acrobat, and smart highlighters (available for sale at *http://firedoodle.com*).
Resources	Various books by Keeley and colleagues to probe student thinking (Keeley 2005, 2008, 2011; Keeley, Eberle, and Farrin 2005; Keeley, Eberle, and Tugel 2007; Keeley, Eberle, and Dorsey 2008; Keeley and Harrington 2010).

Talk About Thinking*

Students externalize their thinking by using a variety of strategies to think out loud. By verbalizing their thinking, students gain awareness of and control over their problem-solving abilities and a fresh perspective on their own thoughts (Hartman and Glasgow 2002).

The Research	Talking about their thinking helps students develop the language of thinking (Blakey and Spence 1990).
Classroom Implications	• You can model and discuss thinking, labeling your thinking as you talk. For example, you can analyze a set of data that students have gathered about growth of a population of mice, talking through what the data indicate. You might say, "I notice the line graph has an upward slope but then it levels off. That tells me that the number of individuals increased for a while but then the population stopped growing. I wonder why that happened?" Use of words like *slope* and *increased* are good insights into your thinking process. Hearing you talk in this way helps students realize just what thinking processes are (Blakey and Spence 1990). • Paired problem solving formalizes this process. Using this technique, a student talks through a problem, describing his or her thinking, while a partner listens and asks questions to clarify thinking (Blakey and Spence 1990). Paired problem solving makes problems more engaging, promotes self-monitoring and self-evaluation, and gives students feedback on their thinking. It also improves collaboration and communication. It is necessary, however, for you to monitor progress of each pair of students and provide them feedback (Hartman and Glasgow 2002). • Reciprocal teaching (Palincsar and Brown 1985) is another activity that formalizes the process of labeling one's thinking. In this case, the teacher and the students take turns assuming the role of the teacher in leading a dialogue regarding segments of a text. The dialogue uses four strategies: summarizing, question generating, clarifying, and predicting (see Resources on the next page for more information on this strategy).

(continued)

CHAPTER 2: Implementation of the Framework

Instructional Tool 2.2 *(continued)*

Application Example	Imagine that a student pair has just completed working through a simulation about day and night. They are trying to develop an explanation of the simulation, based on their current understandings. One student verbalizes all the thoughts she has as she thinks through what happened in the simulation. The other student actively listens. He also points out what he thinks are errors, examines the accuracy of the statements, and probes the "thinker" to continue voicing her thinking. They can then change roles, and the second student voices his thoughts about the simulation and the first student actively listens and probes. At the end of this activity, they prepare a written summary of their explanation.
Technology Applications	There are very simple voice annotations in PowerPoint, as well as in VoiceThread, which is a much more powerful and collaborative tool: *http://voicethread.com*.
Resources	Reciprocal teaching: *www.ncrel.org/sdrs/areas/issues/students/atrisk/at6lk38.htm*

Plan and Self-Regulate*
This strategy involves estimating time requirements, organizing materials, scheduling procedures needed to complete an activity, and developing evaluation criteria (Blakey and Spence 1990).

The Research	• For students to become self-directed, they must take on increasing responsibility for planning and regulating their learning (Blakey and Spence 1990). • Contractual agreements with students in which they set goals and subgoals (within the context of teacher goals) have a positive impact on planning and self-regulating (Marzano, Pickering, and Pollock 2001). • Graphic organizers can act as thinking tools and memory support systems that scaffold self-regulation. They help students to uncover their prior knowledge and can be used to mark progress and to contrast what was known to what is now known (Lipton and Wellman 1998).
Classroom Implications	• Peel the Fruit is a tool to make thinking visible and to plan and track over time the exploration of a topic (Ritchart and Perkins 2008). This routine allows the entire class, small groups, or individuals to track progress on a long-term project. Further information can be found at the Visible Thinking website (*www.pz.harvard.edu/vt*). • Encourage students to personalize the goals you establish and adapt them to their personal needs. Your goals should be general enough to allow this flexibility (Marzano, Pickering, and Pollock 2001). • Use of rubrics is an effective way for students to track their efforts and the impact of those efforts on their achievement (Marzano, Pickering, and Pollock 2001).

Instructional Tool 2.2 *(continued)*

Application Example	Students are working on a long-term research project—for example, the study of a local habitat. They can use a planning map (e.g., Peel the Fruit [*www.pz.harvard.edu/vt*]) initially to map and plan their work, add tasks, establish research routines, and track progress. They periodically revisit the map, choose next steps, and monitor their progress. "'Peeling the fruit' is a metaphor for getting familiar with the surface of something, seeing puzzles and mysteries to investigate, and pursuing these in various ways to arrive at core understandings" (*www.pz.harvard.edu/vt*).
Technology Applications	NoteStar, an online tool, helps students develop research papers. It lets them create subtopics to research, assign topics to group members (if the research paper is being developed by more than one student), take notes, track source information, and organize notes and sources. It is found at *http://notestar.4teachers.org* and can be used alone or together with ThinkTank (*http://thinktank.4teachers.org*), another tool that helps students manage a topic for online research.
Resources	Monitoring and self-management site: *www.muskingum.edu/~cal/database/general*.

Debrief the Thinking Process*

This strategy includes activities that help students bring closure to a lesson and focus discussion on the thinking process itself. It shows students how to develop their awareness of the ways in which various strategies might be used in other situations (Blakey and Spence 1990).

The Research	• When students talk about learning, they can check their thinking and performance, gain deeper understandings of their learning, and use better strategies for planning and monitoring their work. It also prepares students for the risk-taking required of learning (Davies 2003). It increases their confidence in their thinking and willingness to share ideas in a school environment where typically answers are "right" or "wrong." • Sharing their work and organizing evidence of their learning helps students see clearly what they have learned, what they still need to learn, and what kinds of support are required for them to learn. The presence of others students encourages reflection (Davies 2003), which improves the metacognitive skills of the students who are sharing their work.
Classroom Implications	Blakey and Spence (1990) recommend a three-step process. First, you facilitate a review of an activity, eliciting student responses on thinking processes and feelings. Next, your students classify related ideas and identify the thinking strategies they used. Third, they evaluate their success, identify helpful strategies, and eliminate unproductive strategies.
Application Example	You can use the Blakey and Spence (1990) three-step process to have student groups complete an inquiry activity about, for example, states of matter. Review the activity with the class and ask students to summarize their thinking and feelings about the process they have gone through. Then have them complete an entry in their thinking journals about what challenged—and what clarified—their thinking; why they think they were challenged as they were; and how certain ideas clarified their thinking. Finally, ask each group to talk about the thinking processes they used and whether or not they were effective. This discussion is likely to lead the group to revisit the activity and to modify their initial approaches.

(continued)

CHAPTER 2: Implementation of the Framework

Instructional Tool 2.2 *(continued)*

	Self-Evaluate*
	Students are self-evaluating when they judge the quality of their work based on evidence and explicit criteria. The purpose of self-evaluation is to do better work in the future.
The Research	• Self-evaluation enhances self-efficacy and increases intrinsic motivation (Rolheiser and Ross n.d.). • When students set criteria, self-assess, and reset criteria, they better understand assessment and the language of assessment. They acquire a clearer image of what they need to learn, where they are in relationship to that, and how they might take steps to get there (Davies 2003).
Classroom Implications	• We can move students toward self-evaluation by first providing guided self-evaluation—for example, individual conferences with the teacher and use of checklists that focus on thinking processes. We can then slowly allow students to complete this process more independently (Blakey and Spence 1990). • We should constantly encourage students to compare their current thinking to their original thinking and try to determine what helped them achieve their current understandings. Various graphic organizers support this before-and-after process. • Rolheiser and Ross (n.d.) recommend a four-stage process: (1) involve students in defining the criteria used to evaluate performance, (2) teach students how to apply the criteria to their work, (3) give students feedback on their self-evaluations, and (4) help students develop goals and action plans.
Application Example	Students are given time in class to keep a journal in parallel with other learning processes as a means to practice self-evaluation.
Technology Applications	Once again, use of VoiceThread (*http://voicethread.com*) is appropriate.
Resources	*Science Formative Assessment: 75 Practical Strategies for Linking Assessment, Instruction, and Learning* (Keeley 2008) includes numerous strategies for self- and peer-assessment.
	Keep a Thinking Journal* or Learning Log See Instructional Tool 2.4 for more information on these strategies.

Instructional Tool 2.2 *(continued)*

3. Creative Thinking and Learning	
Visible Thinking: Creativity Routines Creativity Routines look at purposes and audiences, generating creative questions, creative thinking about options, creative decision making, and ways to consider various perspectives. Explore the Visible Thinking website (*www.pz.harvard.edu/vt*) to learn more.	
The Research	See Creativity Routines on the Visible Thinking website.
Classroom Implications	These easy-to-learn routines can be incorporated into any classroom and can address various content areas. Simply select a routine that aligns well with the targeted content goals.
Application Example	A great routine to use in a decision-making situation is called Options Diamond (*www.pz.harvard.edu/vt*). Perhaps your students are engaged in a case involving an environmental impact decision. They draw a large diamond. In the center of the diamond they write the decision that has to be made. At the left and right corners of the diamond, they write the one or two main trade-offs of making a particular decision. (As noted on the website, "Usually there are trade-offs or tensions between [two or more options] that make the decision hard: Choose one and you get X but lose Y; choose the other option and you lose X but get Y.") Students brainstorm (1) solutions for each trade-off; (2) compromises between the trade-offs that they write at the bottom point of the diamond; and (3) clever solutions that combine what seem to be the opposites from the right and left corners and write these at the diamond's apex. They then reflect on the diamond to determine what they learned.

Note: Instructional Tool 2.3, "Instructional Strategy Selection Tool," is on pages 42–44.

CHAPTER 2: Implementation of the Framework

Instructional Tool 2.4
Sense-Making Approaches: Linguistic Representations—Writing to Learn

	Learning Logs
	The terms *learning log* and *science notebook* are often used interchangeably. However, learning logs are a less-structured way for students to maintain written records of their observations and thinking. They permit a free-form type of writing that can express feelings as well as observations.
The Research	• Most writing in science classes has involved writing to answer test questions rather than as a form of dialogue between the student and the teacher. In addition, writing is rarely used in science classes as a way for students to express their thinking, either for their own benefit or for sharing with their peers. Thus, students lose potential opportunities to learn more about science and about the writing process (Rangahau 2002). • Writing can be used to clarify ideas in and about science (Hipkins et al. 2002; Rangahau 2002), and journaling and micro-themes can be used to promote students' personalization and understanding of science (Ambron 1987). Learning logs integrate content, process, and personal feelings and help students learn *from* their writing as opposed to writing *about* what they learn (Costa 2008). Extended writing sequences can be used for metacognitive reflection for students about their understandings of concepts, procedures, and the nature of science (NOS) (Baker 2004; Rangahau 2002). If you use open-ended questions as writing prompts, you can assess your students' understanding of content as well as their thinking skills, especially their abilities to analyze, evaluate, and solve problems (Freedman 1994).
Classroom Implications	• Be sure to respond carefully to your students' writing. When they know that you act on what they write and support them in the process, they are better able to identify their learning difficulties and overcome them (Rangahau 2002). Be careful in how you provide feedback. Rather than giving feedback in the form of a grade, indicate how students might improve their work. A constructive comment is more effective than praise. It is helpful to provide suggestions for what to do next or ask questions that prompt students to connect what they learn to other experiences (Harlen 2000). • Learning logs can engage student thinking about a topic because they require students to focus on content in their writing (Barton and Jordan 2001). A journal or learning log allows your students to reflect on their thinking in a diary format. It lets them make notes about their awareness of difficulties and comment on steps they've taken to deal with these difficulties (Blakey and Spence 1990). You can use open-ended questions developed and used as probes to assess students' thinking and content understanding. Consider using the wonderful probes that make up the *Uncovering Student Ideas in Science* series (see Resources on the next page).

Instructional Tool 2.4 *(continued)*

Application Example	A teacher is teaching a unit on animal adaptations. She asks an open-ended question: "What is it about a dog's fur that might help keep it warm?" Students are asked to provide a written response in their learning logs and are given at least five minutes to write this response. The teacher gathers the students' logs at the close of the day and reviews the students' writing. She writes comments that include questions such as, "What convinces you that this is the case?" "How would you test this idea?" and "How will you find out more about this topic to make a stronger response?" The following day, students review their logs and respond in writing to any questions the teacher has asked. Small groups read one another's responses and discuss the various explanations, reaching some consensus on their thinking.
Technology Applications	• Students can keep electronic learning logs and share them with other students and the teacher, allowing them to reflect on their own thinking as well as defend their viewpoints and react to those of others (Baker 2004). • Blogs and vlogs are very appropriate to use here.
Resources	• *Open-Ended Questioning: A Handbook for Educators* (Freedman 1994) • *Teaching, Learning & Assessing Science 5–12* (Harlen 2000) • Various books to probe student thinking have been written by Keeley and colleagues (Keeley 2005, 2011; Keeley, Eberle, and Farrin 2005; Keeley, Eberle, and Tugel 2007; Keeley, Eberle, and Dorsey 2008; Keeley and Tugel 2009; Keeley and Harrington 2010). Read more about the *Uncovering Student Ideas in Science* series at www.nsta.org/publications/press/uncovering.aspx.

Probes	
	Probes are formative assessments used for both diagnostic and monitoring purposes and are considered assessments *for* learning. The purpose of the probes is to reveal to you and your students the conceptions they have about various science concepts. We focus here specifically on the probes developed by Page Keeley and her colleagues (see Resources, p. 92).
The Research	Information gained from use of the probes has generated much feedback and general research has been conducted. Here are some key findings: • Probes help teachers know more about what their students understand. • Probes significantly change instruction in the classroom. • Teachers tend to listen very closely to their students' reactions to the probe, which helps the teachers determine their next instructional steps. • Probes are used successfully in professional development, helping teachers learn more about content (teachers confront their own misconceptions) and instruction (they discover that some traditional science activities actually reinforce common misconceptions).

(continued)

Instructional Tool 2.4 *(continued)*

Classroom Implications	• It is important to understand what each probe is designed to reveal so that you are sure to select probes that align with your learning target. • You might use multiple probes that target a single concept, which will give you even more information about students' understandings and about how to modify instruction. • In the Page Keeley et al. books, be sure to read the Explanation section in the Teacher Notes for each probe. It provides clarification about the content. • Carefully consider the Curricular and Instructional Considerations section to assure yourself that you are using a probe that is appropriate for your grade level. Also use the Administering the Probe section for ideas about using the probe appropriately at your grade level. • Carefully read the Related Research section for each probe. You will better understand the purpose of the probe and anticipate possible responses from your students.
Application Example	Instead of the card sort explained in Chapter 2 of this book (see pg. 67), use the probe "Is It Food?" (Keeley and Tugel 2009, p. 90). This probe lists different items, including some that are food (e.g., milk, bread, turkey) and some that are not food (e.g., salt, water). Students check off which items they think fit the scientific definition of *food* and then are directed to "Explain your thinking. What definition or 'rule' did you use to decide if something can scientifically be called *food*?" This is an all-purpose prompt that can be used *well before* instruction, as you *begin* instruction, and *during* instruction to determine your class's current level of understanding.
Technology Applications	• Students could complete a probe at home and then blog or vlog with their peers. • Students' probe responses can be kept in their electronic journals, which can be made available for your feedback.
Resources	Use the *Uncovering Student Ideas in Science* series by Page Keeley and her colleagues. Go to *www.nsta.org/publications/press/uncovering.aspx* for more information.

Instructional Tool 2.4 *(continued)*

	Science Notebooks
	Science notebooks are a stable location for students' work, a record of the information they value, and a window into their mental activities. They are a link between science and literacy and include writing, graphs, tables, and/or drawings that help students make meaning of science learning experiences. Science notebooks focus on the more structured writing of science while science journals or learning logs are more free-form. Although both types of writing are important, science journals and science notebooks should be separately maintained (Hargrove and Nesbit 2003).
The Research	• Most writing in science classes has involved writing to answer test questions rather than as a form of dialogue between the student and the teacher. In addition, writing is rarely used in science classes as a way for students to express their thinking, either for their own benefit or for sharing with their peers. Thus, students lose potential opportunities to learn more about science and about the writing process (Rangahau 2002). • Writing can be used to clarify ideas in and about science (Hipkins et al. 2002; Rangahau 2002). The first goal of writing in science is to understand something because writing is a tool with which to think. When students use their science notebooks during classroom discussions, it helps them construct meaning from the science phenomena they observe (Harlen 2001). Science notebooks focus on making sense of phenomena under investigation and help students provide evidence when they are answering the questions of their teacher or their peers. • Science notebooks are useful for assessment when (1) most of the work in the notebook is narrative and centered around authentic science tasks, (2) the notebook work is purposeful, with students investigating their own questions, (3) "right" answers or conclusions are uncommon, and (4) the notebook provides information not only to the teacher, but also to the student and possibly to parents (Hargrove and Nesbit 2003).
Classroom Implications	• There are various ways to structure a science notebook. Consider breaking it down into these seven parts: focus question, prediction/hypothesis, planning, data, claims and evidence, making meaning conference, conclusion/reflection. Another format might include the following components: recording and organizing data, technical drawings, and students' questions (Campbell and Fulton 2003). • Science notebooks are valuable in connecting what should be taught (the standards) to classroom instruction and to the actual content students learn. Make sure to respond carefully to your students' writing. When they know that you act on what they write and support them in the process, they are better able to identify their learning difficulties and overcome them (Rangahau 2002). Be careful in how you provide feedback. Rather than giving feedback in the form of a grade, indicate to students how they might improve their work. A constructive comment is more effective than praise. It is helpful to provide suggestions for what to do next or ask questions that prompt students to connect what they learn to other experiences (Harlen 2000). This can be done through written or oral responses. Timeliness of feedback is essential. If you delay providing feedback, student misconceptions may persist.

(continued)

Instructional Tool 2.4 *(continued)*

Application Example	Students maintain science notebooks during their study of the flow of matter and energy in ecosystems (see Chapter 2).
Technology Applications	Blogs, wikis, and online science notebooks are good applications.
Resources	• www.sciencenotebooks.org • *Science Notebooks: Writing About Inquiry* (Campbell and Fulton 2003)

Writing Scientific Explanations	
Scientific explanations address a combination of the goals of explanation (how and why something happens) and argumentation (written or oral social activity that justifies or defends a standpoint for an audience). Students justify their explanations of phenomena by making claims and supporting those claims with appropriate evidence and reasoning (McNeill and Krajcik 2006).	
The Research	• "Transmissive" writing is a type of writing that transmits information and is often seen in science classrooms. However, if transmissive writing is used exclusively, students will not have the opportunity to use writing to develop and understand the nature of science; in particular, they will not be able to use writing to outline evidence for an argument and to explain the reasoning behind planning decisions (Rowell 1997). Supporting students in the construction of explanations results in improved student understandings about inquiry, science content, and science literacy (McNeill and Krajcik 2008). • It is difficult for students to learn how to construct explanations while they are in the midst of learning content in a particular area, so teacher support is essential. Furthermore, context-specific scaffolds are more effective than generic explanation scaffolds in promoting students' abilities to write scientific explanations and explain their understandings of science content (McNeill and Krajcik 2006).

Instructional Tool 2.4 *(continued)*

Classroom Implications	• Learning to construct explanations is hard for students, so science teachers should provide them with a framework such as the following: (1) make a claim, (2) provide evidence for the claim, and (3) provide reasoning that links the claim to the evidence. • Actions the teacher can take to help students construct scientific explanations include the following: (1) uncover students' prior knowledge about what *explanation* means, (2) generate criteria for explanations, (3) make the framework explicit, (4) model the construction of explanations, (5) provide students with practice opportunities, (6) practice the critique of explanations, (7) give feedback to students, and (8) provide opportunities for students to revise their work. Students can first practice writing scientific explanations for phenomena with which they are already familiar and for which their content knowledge is well formed. They can then move to writing scientific explanations related to their own investigations. • You can support students' practice in writing explanations by using a "summary frame" with them before they start to develop explanations on their own. Summary frames are series of questions you provide students, centered on a specific type of information. The questions highlight the important elements of that type of information and are intended to help students write summaries of the information (Marzano, Pickering and Pollock 2001). When writing scientific explanations, you employ the "argumentation pattern," using text that tries to support a claim. The argumentation pattern contains these elements: evidence, claim, support, and qualifier (restriction on the claim or evidence that counters the claim). Specific questions might be (1) What information is provided (in the text) that leads to the claim? (2) What claim does the author make about the problem or situation? (3) What examples does the author use to support the claim? (4) What concessions does the author make about the claim? Because reading teachers commonly use this approach, you might be familiar with it. If so, that should make it easier when you use it to teach students about writing explanations. Students can then use this approach when writing explanations about their own investigations.
Application Example	Have students describe everyday instances where they might need to explain something—how to use their Wii, how to play a new game they got for their birthday, how to do a particular math problem, or what to tell their parents if they get home late from school. This activity relies on their everyday use of the word *explain*. Share explanations with the class and analyze these everyday explanations. Which require facts or other evidence? How does the explanation usually begin? Which are "good" explanations and which are not? During this conversation, make sure students distinguish between opinion-based and evidence-based explanations. Notice that this activity addresses supporting actions #1 and #2 described in the second paragraph in Classroom Implications, above.

(continued)

Instructional Tool 2.4 *(continued)*

Technology Applications	Blogs, wikis, and online lab notebooks are good applications.
Resources	• *Questions, Claims, and Evidence: The Important Place of Argument in Children's Science Writing* (Hand et al. 2008) • "Inquiry and Scientific Explanations: Helping Students Use Evidence and Reasoning" (McNeill and Krajcik 2008) • Various books to probe student thinking have been written by Keeley and colleagues (Keeley 2005, 2011; Keeley, Eberle, and Farrin 2005; Keeley, Eberle, and Tugel 2007; Keeley, Eberle, and Dorsey 2008; Keeley and Tugel 2009; Keeley and Harrington 2010). Read more about the *Uncovering Student Ideas in Science* series at *www.nsta.org/publications/press/uncovering.aspx*. • *Everyday Science Mysteries* series (Konicek-Moran 2008, 2009, 2010, 2011) • A nice lesson is found at *www.readwritethink.org/lessons/lesson_view.asp?id=872*.
	**Science Writing Heuristic** The science writing heuristic (SWH) is a structured approach that combines guided inquiry, collaborative group work, and writing-to-learn activities. "The SWH provides an alternate format for students to guide their peer discussions and their thinking and writing about how hands-on guided inquiry activities relate to their own prior knowledge via beginning questions, claims and evidence, and final reflections" (Burke et al. 2005, p. 2).
The Research	Use of SWH both independently and in connection with textbook strategies had a positive impact on student conceptual understanding and metacognition (Wallace, Hand, and Yang 2004). The SWH process has been used in a variety of classrooms across grade levels and science disciplines. Productive integration of SWH has produced student-learning gains (Burke, Greenbowe, and Hand 2005).
Classroom Implications	The SWH process is highly effective, but also complex. The process engages students in activities to identify preinstructional ideas, prelaboratory activities to begin thinking about the concepts, the laboratory activity, small-group sharing and comparing of data, comparing ideas to those in textbooks or other print resources, individual reflection and writing, and exploring postinstruction understanding. Each component is essential and requires attention by the teacher. Note that this is a process that entails multiple strategies from across the Instructional Tools in this chapter. Instructor preparation is essential for the successful integration of SWH into a science course and for positive impacts on student learning (Burke, Greenbowe, and Hand 2005). This strategy is included here primarily to encourage you to consider training in SWH. We encourage you, if you have the opportunity, to attend a professional development session in your school district or state or at an NSTA conference. Once you are familiar with SWH, you can use it in your lesson development.

Instructional Tool 2.4 *(continued)*

Application Example	Because of the complexity of SWH, it would take several pages to thoroughly describe an application example here.
Technology Applications	Blogs, wikis, and online lab notebooks are good applications.
Resources	• A brief overview of SWH is found at *http://edutechwiki.unige.ch/en/Science_writing_heuristic*. • You can download a template at *www.aea11.k12.ia.us/science/Heuristic.html*.

CHAPTER 2: Implementation of the Framework

Instructional Tool 2.5

Sense-Making Approaches: Linguistic Representations—Reading to Learn

	Vocabulary Development These strategies help students understand words in a lesson or unit that are unique to science. The strategies also identify certain words that have different meanings in science than they do in everyday use (e.g., *fault, force, food*).
The Research	• Students' gradual understanding of new vocabulary improves comprehension of text, regardless of age or population (Magnusson and Palinscar 2004). Furthermore, coaching in the language of science (both vocabulary and grammar) simultaneously supports reading and science literacy (Hipkins et al. 2002; Rangahau 2002). However, it is important to make sure students understand the conceptual foundation for a topic before introducing specialist vocabulary (Hipkins et al. 2002). • Words with dual meanings (e.g., *fault, force, food*) cause more problems in science lessons than words with single meanings. Often in science, nouns can be substituted for verbs or an entire sequence of events (e.g., *evaporation*), and nouns can be used as adjectives (e.g., *elephant population*). Explicit coaching about these grammatical features of science enhances reading literacy and science literacy (Rangahau 2002). • Learning science vocabulary is important so that students can learn the discourse of science. Embedding the language of science in guided-inquiry experiences allows you to model how science language is used (Magnusson and Palinscar 2004). Words used for hypothesizing, comparing, and other aspects of science reasoning (e.g., *frequently* or *simultaneously*) cause difficulty. Giving your students practice with argumentation (written and oral) helps them build understandings of these words (Rangahau 2002). • A good portion of vocabulary learning should occur as students learn subject matter because the context in which vocabulary is learned is important (Magnusson and Palinscar 2004). Use of models and analogies builds shared meanings that facilitate the communication of individual understandings of science technical vocabulary (Rangahau 2002).
Classroom Implications	• Vocabulary instruction not only prepares students to read and understand science text but also develops students' conceptual understanding so they can communicate new ideas. You might find it a challenge to prepare for vocabulary instruction. Four steps that help in this process are (1) identify learning goals, (2) develop a vocabulary list based on the goals, (3) determine required levels of understanding for the terms, and (4) select appropriate strategies (Barton and Jordan 2001). • One strategy that helps students with terms that don't require in-depth understanding is the Student VOC (Vocabulary) Strategy. Strategies that work better for terms that require deeper exploration include concept definition mapping, the Frayer model, and semantic feature analysis. These strategies can be used throughout a unit, with modifications as understanding grows (Barton and Jordan 2001). The Just Read Now website explores these strategies (see Resources, p. 99). • It is important to offer multiple opportunities to learn new terms in context, provide some instruction on the concepts prior to reading, help students connect an image to a term, and focus only on those terms critical to learning the new content (Barton and Jordan 2001).

Instructional Tool 2.5 *(continued)*

Application Example	In a lesson on food chains, a teacher found that his students did not fully understand the difference between plants and animals and some students did not understand the differences among living, dead, and nonliving. He used semantic feature analysis (i.e., using a grid to explore how sets of things are related to one another) with these sets of essential terms. However, there were also terms required by his school district that were not essential to students' conceptual understanding of the topic. He decided to use the Student VOC (Vocabulary) Strategy for these terms because in-depth understanding was not essential.
Technology Applications	Inspiration software can be used for semantic mapping.
Resources	• Seeds of Science: Roots of Reading website: www.seedsofscience.org/strategyguides.html • Science portion of Just Read Now website: www.justreadnow.com/content/science/index.htm • Connecting Elementary Science and Literacy website (connects literacy to inquiry): http://cse.edc.org/products/scienceliteracy/matrixhome.asp • *Teaching Reading in Science: A Supplement to the Second Edition of Teaching Reading in the Content Areas Teacher's Manual* (Barton and Jordan 2001
Informational Text Strategies	
Informational text strategies help students better understand text structure, text coherence, and appropriateness for a particular audience and, as a result, have a positive impact on the learning of science.	
The Research	• Reading and comprehension skills are important in science because of specialist language and grammatical features. This may require significant teacher support (Rangahau 2002). You should help students recognize the role of their prior knowledge and teach them how to use that knowledge when they learn science through reading (Barton and Jordan 2001). • Use of text in inquiry-based instruction is important because learning from text is a standard science practice and a way to promote both text comprehension and science instruction (Magnusson and Palinscar 2004). Hands-on activities stimulate student questioning, and the search for answers in textbooks or other science materials provides frames of reference to construct meaning from text (Nelson-Herber 1986). • Reading and inquiry both require that learners are aware of and use the appropriate discourse structures, can coordinate information across texts, and interpret multiple representations in texts (Magnusson and Palinscar 2004). Explicit instruction with students to both recognize and represent common text structures significantly improves student learning, as does familiarizing students with the different ways that information is presented in various text forms (Barton and Jordan 2001).

(continued)

CHAPTER 2: Implementation of the Framework

Instructional Tool 2.5 *(continued)*

| **Classroom Implications** | • At the beginning of the year, walk students through a section of the textbook. They need to understand text presentation—that is, how the material is laid out; visual clues, such as illustrations and graphs; and textual clues such as headings, lists, and titles of visual elements (Vasquez, Comer, and Troutman 2010). Model for students how to predict text content based on these presentation clues.
• If you are working with a textbook for the first time, before the school year starts you should assess the textbook for coherence. If there are limitations in the presentation, you need to pre-identify main ideas and then, as you work through the textbook with your class, make certain that students understand how concepts in paragraphs are related. Many science textbooks are not "user friendly" for students in elementary school. If that is the case, you may need to use alternative materials until appropriate textbooks are adopted (Barton and Jordan 2001).
• The language used in books and other materials about science (i.e., science "text") is specific to science. Ways that science texts differ include different kinds of nouns and referent items (*generic* rather than *particular*), verb tense (use of the present tense instead of the past tense, which young children are used to from storybooks), and vocabulary (see Vocabulary Development, pp. 98–99). Science texts also include a range of images and designs (Varelas and Pappas 2006).
• Text structures often used in science with which students should become familiar include the following: comparison/contrast, concept definition, description, generalization/principle, and process/cause and effect. Teachers should scaffold students' learning about each structure when students begin to learn a skill or concept and use words that will be part of the content focus (Barton and Jordan 2001).
• Reading strategies that can be applied across disciplines in an elementary classroom include discussion strategies (e.g., creative debate discussion webs [see www.justreadnow.com/strategies/debate.htm], Question-Answer Relationships (QAR), and Think Pair Share); active reading strategies (e.g., anticipation guides, KWL, and reciprocal teaching), and organization strategies (concept diagrams, graphic organizers, and two-column notes). Several of these strategies also help identify student preconceptions, which helps in teacher planning and allows students to grapple with their own thinking. Anticipation guides, the Directed Reading Thinking Activity (DR/TA), various graphic organizers, Group Summarizing, KWL, PLAN, Problematic Situation, Proposition/Support Outline, reciprocal teaching, and Think-Aloud are all effective strategies to help students make meaning from text (Barton and Jordan 2001). Online resources outline many of these strategies (see Resources on next page). |

CHAPTER 2: Implementation of the Framework

Instructional Tool 2.5 *(continued)*

Application Example	Students have completed an activity that shows forces causing changes in direction of movement, and they have developed explanations for what happened in the activity. The teacher then uses Pairs Read as a strategy. She provides pairs of students with a passage from the textbook or other resources accessible to readers in fifth grade. Each pair uses a different resource. She tells them that one student in each pair will be the coach and the other will be the reader. The reader reads the first paragraph aloud to the coach, and then the coach summarizes the paragraph. They then reverse roles. If the selection is more than two paragraphs, they continue to rotate being the reader and the coach. Each pair of students then summarizes what they read, and various pairs share their summaries with the class.
Technology Applications	NA
Resources	• Seeds of Science: Roots of Reading website: www.seedsofscience.org/strategyguides.html • Visit www.mcrel.org to find sites that offer suggestions for science resources that supplement text. • Science portion of the Just Read Now website: www.justreadnow.com/content/science/index.htm • NSTA annually publishes "Outstanding Science Trade Books for Students, K–12": www.nsta.org/publications/ostb. • *Teaching Reading in Science: A Supplement to the Second Edition of Teaching Reading in the Content Areas Teacher's Manual* (Barton and Jordan 2001) • *Strategies That Work: Teaching Comprehension for Understanding and Engagement* (Harvey and Goudvis 2007) • *The Comprehension ToolKit: Language and Lessons for Active Literacy* (Harvey and Goudvis 2011)
Reflection Strategies	
Reflection strategies develop students' metacognitive abilities and are important if students are to become effective readers. Specific strategies can be used to promote reflection on reading in science.	
The Research	Reflection strategies that enhance metacognition related to reading in science help students (1) better understand science text, science reading, and science reading strategies; (2) enhance skills required to read science text and use science reading strategies; and (3) understand why and when to use particular strategies. These abilities help students (1) better understand science text, science reading, and science reading strategies; (2) improve the skills required to read science text and use science reading strategies; and (3) understand why and when to use particular strategies (Barton and Jordan 2001).

(continued)

CHAPTER 2: Implementation of the Framework

Instructional Tool 2.5 *(continued)*

Classroom Implications	Reflective questioning, reflective writing, and discussion enhance students' metacognition as related to learning science (including reading science) and enhance students' understanding of science content. Various reflection strategies include creative debate; discussion webs; learning logs; Question-Answer Relationships (QAR); Questioning the Author (QtA); Role/Audience, Format/Topic (RAFT); and Scored Discussion (Barton and Jordan 2001). See Resources, p. 102, for learning more about these strategies.
Application Example	Imagine that students are learning about states of matter. The teacher asks students to write responses in their learning logs about the following question: "Why are there bubbles in a pan of boiling water?" She provides time for students to think about their responses and they then write down the responses. Next, the teacher teaches a lesson that focuses on the idea that water vapor is a gas. The teacher then asks the students to revisit their responses in their learning logs and reflect on how their ideas have changed.
Technology Applications	Online journals and discussions can be implemented. Consider using a classroom blog where scientific discussions can occur.
Resources	• *Teaching Reading in Science: A Supplement to the Second Edition of Teaching Reading in the Content Areas Teacher's Manual* (Barton and Jordan 2001) • Science portion of the Just Read Now website: *www.justreadnow.com/content/science/index.htm*

Thinking-Process Maps

Thinking-process maps help students who are reading and interpreting information that is either science-content specific or interdisciplinary. For more information about the many other uses of thinking-process maps, see Instructional Tool 2.8, p. 114, and visit Thinking Maps at *www.thinkingmaps.com*.

Instructional Tool 2.6

Sense-Making Approaches: Linguistic Representations—Speaking to Learn

Large- and Small-Group Discourse Discourse is used to make sense of science learning experiences. The teacher and students explore ideas, pose questions, and listen to multiple points of view to establish understanding (Mortimer and Scott, 2003). Discourse benefits student learning, teacher-student rapport, equity, expectations, and formative assessment (ASCD 2008).	
The Research	• Students can engage freely in focused talk in science without being taught rules, conventions, structures, or vocabulary, especially if realistic and authentic problems are the focus of the discussions. You can develop focus using strategies that present authentic problems, focus on scientific issues, and lead to productive follow-up activities. Better discussions occur when students see several plausible viewpoints, resulting in students' own theory determination. When only one viewpoint is presented, there is no reason to have a discussion. Alternative viewpoints are effectively presented via concept cartoons or puppets (Keogh and Naylor 2007). Discussion based on reading materials such as magazine articles, online commentary, and books also enriches the discourse. Examination and discussion of text promotes skills of analytical reading, careful listening, citing evidence, respectful disagreement, and open-mindedness (Hale and City 2006). • "Accountable talk" deepens conversation and understanding of the topic studied (ASCD 2008). Norms and skills for "accountable talk" must be explicitly taught. • Large- and small-group discourse promotes learning about concepts, metacognition, and the nature of science; it also develops positive attitudes. Whole-class discussion works best in the context of an activity because it provides shared experiences for students (Hipkins et al. 2002). Small-group discourse lets students build on one another's ideas and generate explanations. Whole-class discourse demands more clarity and explanatory power on the part of students. If students understand the purposes of the two types of discourse, they will better develop shared understandings (Woodruff and Meyer 1997). There should be a balance between whole-class and small-group discussions. Notice that they reflect the two types of scientific discourse, one within the laboratory and one among laboratories working on similar research (Cartier 2000; Hipkins et al. 2002).

(continued)

Instructional Tool 2.6 *(continued)*

Classroom Implications	• Purposes of discussion during inquiry include eliciting ideas, planning investigations, and developing understanding. You can use surveys or questioning to assess students' preinstructional ideas (Hartman and Glasgow 2002). Let your students talk in their own registers (not formally). Give them time and space to work productively without intruding (Keogh and Naylor 2007). • You can encourage deeper discussion and investigation with student-generated questions, open-ended questions, student choice of inquiry topics, and more time for student research and exploration (ASCD 2008). Use strategies like card sorts, concept cartoons, odd one out, and graphic organizers to effectively engage students and promote student self-motivation and self-sustaining conversation. Concept cartoons can provide a safe environment and focus the discussion during argumentation (Keogh and Naylor 1999, 2007). (See Instructional Tool 2.9, p. 121, for further information on concept cartoons.) • What is your role during small-group discussions? You should provide students with carefully selected materials to focus discussion and decision making, require them to operate at high cognitive levels, ensure that everyone has the opportunity to speak, and require student products that result from discussion (Hipkins et al. 2002). Avoid individual student worksheets because they often inhibit student talk; students tend to complete the worksheets instead of discussing (Keogh and Naylor 2007). Your questioning is essential when addressing a topic that requires thought and deduction. Use cognitive, speculative, affective, and management questions, each asked at various levels (Hartman and Glasgow 2002). Avoid questions that elicit factual recall because such questions bring discussion to a halt. It is more productive to use questions that probe thinking, build problem-solving steps, encourage participation, and sequence conversation (ASCD 2008; Hipkins et al. 2002). • When forming groups, keep in mind that role allocation in mixed-gender groups promotes more conceptual dialogue; role allocation is less necessary in same-gender groups (Hipkins et al. 2002). You can promote accountable talk by pushing students to (1) clarify and explain, (2) require justification for proposals, (3) recognize and challenge misconceptions, (4) demand evidence from peers for claims and arguments, and (5) interpret and use one another's statements (ASCD 2008).
Application Example	Taking turns, each student is responsible for facilitating large- and small-group discussions that have a variety of goals and outcomes and occur at multiple points in the learning cycle.
Technology Applications	Voice annotations in PowerPoint are simple applications. More powerful and collaborative options are available with VoiceThread (*http://voicethread.com*).

Instructional Tool 2.6 *(continued)*

Resources	• *Quality Questioning: Research-Based Practice to Engage Every Learner* (Walsh and Sattes 2005) • *Open-Ended Questioning: A Handbook for Educators* (Freedman 1994) • *The Teacher's Guide to Leading Student-Centered Discussions: Talking About Texts in the Classroom* (Hale and City 2006) • To see discourse routines using science texts, visit the Seeds of Science: Roots of Reading website: *www.seedsofscience.org/strategyguides.html*.
Student Questioning Student questioning includes not only questioning during small- and large-group discussions but also questioning as an essential feature of inquiry.	
The Research	• Formulating questions is the key to metacognitive knowledge, including strategic knowledge, knowledge about cognitive tasks, and self-knowledge (Walsh and Sattes 2005). Student-generated questions, open-ended questions, student choice of inquiry topics, and more time for student research and exploration encourage deeper discussion and investigation (ASCD 2008). Helping students become aware of what constitutes a good question makes them better learners and improves their questioning abilities (Hartman and Glasgow 2002). • See Instructional Tool 2.1 for more information about student questioning.
Classroom Implications	• Provide your students with models of good questions and discuss effective questioning strategies. Also give them opportunities to practice questioning and provide them with feedback on their questions (Hartman and Glasgow 2002; Walsh and Sattes 2005). • Consider using reciprocal teaching and paired-problem-solving strategies because they require students to ask and answer questions and thus promote questioning abilities and metacogntive knowledge (Walsh and Sattes 2005). • See Instructional Tool 2.1 for more information on student questioning.
Application Example	Students generate questions prior to a class discussion based on their current understandings about the content. They share questions in small groups, and peers critique the questions based on previously learned criteria. Groups then share a single question with the whole class, and the teacher facilitates a class critique of the questions.

(continued)

CHAPTER 2: Implementation of the Framework

Instructional Tool 2.6 *(continued)*

Technology Applications	• Inspiration software can be used to brainstorm/generate questions related to content investigations. This activity, coupled with analysis of each question using the questioning tree (see Resources below), can help students quickly generate questions and then reduce the questions to those that are engaging and investigable. In addition, those questions that are informative and perhaps foundational to the content, but not investigable, might be answered via a web search. This can be done individually, in small groups, or as a whole class. • E-mail/webconferencing/blogging with scientists and other students could be used to generate questions. • Microblogging, like Twitter (*http://twitter.com*), can be used to capture students' questions.
Resources	• The SCENE website (*http://scene.asu.edu/habitat/inquiry.html#4*) offers a nice overview of a questioning cycle in inquiry-based science and a questioning tree (Figure 3 on that site) that guides thinking about questions that are both engaging and investigable. • Two resources that should help support quality questioning, for teacher use and to model for students, are *Quality Questioning: Research-Based Practice to Engage Every Learner* (Walsh and Sattes 2005) and *Open-Ended Questioning: A Handbook for Educators* (Freedman 1994). Walsh and Sattes include a chapter with specific suggestions on how to teach students to generate questions.

CHAPTER 2: Implementation of the Framework

Instructional Tool 2.7

Sense-Making Approaches: Nonlinguistic Representations—Six Kinds of Models

1. Mathematical Models	
Mathematical models can be used to model events, objects, and relationships. Using mathematical models includes learning to abstractly represent things, logically manipulate the abstract representations, interpret the results, and determine the model's appropriateness (AAAS 2001b). Mathematical models come in various forms that include formulas, equations, figures, graphs, and pictograms (Gilbert and Ireton 2003).	
The Research	• Intermediate-level students should learn that they can represent mathematical ideas concretely, symbolically, or graphically. Students often struggle to understand how symbols are used in algebra. They also fail to view the equal sign of equations as a symbol of equivalence between the two sides of the equation and, instead, think it means it is time to begin calculations (AAAS 2001b). • Students of all ages often interpret graphs as literal pictures rather than symbolic representations, but little is known about how graphic skills are learned or about how graph production is related to graph interpretation. It is known than MBLs (microcomputer-based laboratories) improve students' abilities to interpret graphs (AAAS 2001b).
Classroom Implications	You should regularly and frequently involve your students in creating graphs, and you should do so purposefully as models. It is just as important to have them explain their models as to make them so that you can see that they understand what their abstract model—a graph—represents in concrete terms (Gilbert and Ireton 2003).
Application Example	Any formulas and equations you use with your students represent a concept. Be sure to spend time with students helping them interpret what the mathematical model represents.
Technology Applications	Microcomputer-based laboratories (MBLs) let students conduct experiments and generate graphs representing results, spending their time on what the representation means rather than on the creation of the representation itself.
Resources	• *Math: Stop Faking It!* (Robertson 2006) • The National Library of Virtual Manipulatives (*http://nlvm.usu.edu/en/nav/vLibrary.html*)

(continued)

CHAPTER 2: Implementation of the Framework

Instructional Tool 2.7 *(continued)*

2. Physical Models	
Physical models are concrete models that are either two-dimensional (such as diagrams found in textbooks) or three-dimensional (such as working models and scaled models). When students explain their mental models in public, those models become "expressed models."	
The Research	• Generating a concrete representation creates an "image" of the knowledge in a student's mind. Physical models, which are one form of concrete representations, enhance nonlinguistic representations and students' understanding of content (Marzano, Pickering, and Pollock 2001). • If conceptual and metacognitive learning are linked, students' attitudes as they work with physical models improve (Hipkins et al. 2002). When students develop their own models, they are likely to see connections between ideas; student-developed models also help teachers recognize gaps in student understanding. They can be shared and critiqued, as well as changed, as students learn more (Windschitl 2008). • The incorrect use of models can lead to misconceptions. Textbook diagrams can be misleading because they depict in two dimensions something that is actually three-dimensional (Andersson 1990). Furthermore, students tend to see models as concrete mini-representations of larger concrete objects, not understanding the abstract concepts they represent (Grosslight et al. 1991).
Classroom Implications	Student practice in construction, critique, and use of their own models enhances conceptual development (Hipkins et al. 2002). It is important to have students compare various models to the actual concept they represent, clarifying the similarities and differences between the model and the target concept (Gilbert and Ireton 2003).
Application Example	Bottle Biology TerrAqua Columns (Ingram 1993) are one of many physical models for ecosystems. Use these models in the classroom prior to and/or in conjunction with fieldwork. Be certain to discuss with students the strengths and deficiencies of the models as compared to actual ecosystems.
Technology Applications	NA
Resources	*Understanding Models in Earth and Space Science* (Gilbert and Ireton 2003)

Instructional Tool 2.7 *(continued)*

\multicolumn{2}{c}{**3. Verbal Models: Analogies**}
\multicolumn{2}{l}{Creating analogies as a process identifies relationships between pairs of concepts, identifying relationships between relationships (Marzano, Pickering, and Pollock 2001).}
The Research	• Creating analogies is an effective way to identify similarities and differences. They help students see that things that seem to be dissimilar are also similar in some ways. This insight increases understanding of new information and improves student achievement (Marzano, Pickering, and Pollock 2001). Using analogies may also be motivational because they tend to provoke student interest (Cawelti 1999). • Analogies help students construct more accurate conceptions of complex ideas (Hartman and Glasgow 2002), especially when students have alternative conceptions (Cawelti 1999). They help familiarize students with concepts that are outside their previous experiences (Cawelti 1999). • Using multiple analogies in a bridging sequence can help students make sense of initially counterintuitive ideas (Cawelti 1999).
Classroom Implications	• You can use analogy creation to help students identify similarities and differences, a prime strategy to improve achievement. Creating analogies is effective both when you explicitly guide analogy creation and when students independently create their own analogies. Direct instruction is more effective if there are specific similarities and differences on which you want your students to focus. On the other hand, if divergence in student thinking is your goal, have students work independently (Marzano, Pickering, and Pollock 2001). • Finding good analogies to develop conceptual understanding is more difficult when relationships are abstract. You can make ideas more concrete by helping your students form concrete mental images based on personal life experiences and then having them create their own analogies (Hartman and Glasgow 2002). • To be effective, analogies must be familiar to students so they can determine if the features and functions of the analogies are congruent with the targeted concept. Spend time with students discussing the similarities and differences between the analogy and the target. For your students to fully understand the effectiveness of an analogy, it might be helpful to compare multiple analogies to a single learning target (Cawelti 1999). • Another reason that the framework used for an analogy has to be familiar to students is that, otherwise, analogies can be biased socially, experientially, or culturally (e.g., the analogy of cell to city certainly makes more sense to students who are city dwellers) (Hartman and Glasgow 2002). • Discussion of analogies helps students build understanding (Cawelti 1999). You can provide students with an analogy to a concept and have them discuss its relevance and limitations (Hartman and Glasgow 2002).

(continued)

Instructional Tool 2.7 *(continued)*

Application Example	Marzano, Pickering, and Pollock (2001) share a teacher-directed analogy that gives structure to students as they learn to use analogies: "thermometer is to temperature as odometer is to distance" (p. 26). They then suggest that the teacher ask the students to explain how the two relationships are similar. Try this in your classroom. After helping students with several additional analogies, have students use an analog graphic organizer to develop analogies related to the current topic of study in your classroom. Make sure to discuss the strengths and weaknesses of developed analogies. (*Note:* Analog organizers can be found online. One example is at TeacherVision [*www.teachervision.fen.com/graphic-organizers/printable/48386.html*]. For more, google "graphic organizer and analogies.")
Technology Applications	If the analogy is supported by images, standard graphics programs can be used.
Resources	• *Teaching-with-Analogies Model* (*www.coe.uga.edu/twa/PDF/Glynn_2007_article.pdf*) • *Using Analogies in Middle and Secondary Science Classrooms: The FAR Guide–An Interesting Way to Teach with Analogies* (Harrison and Coll 2008) • The Just Read Now website (*www.justreadnow.com/strategies/analogy.htm*) describes and gives examples of word analogies. • *Metaphors and Analogies: Power Tools for Teaching Any Subject* (Wormeli 2009); online at *http://pwoessner.com/2009/12/17/metaphors-and-analogies-power-tools-for-teaching-any-subject*

Instructional Tool 2.7 *(continued)*

4. Verbal Models: Metaphors	
Creating metaphors is a process that identifies a general or basic pattern in a topic and finds another topic that seems different but has the same general pattern. The relationship between the two items in the metaphor is abstract (Marzano, Pickering, and Pollock 2001).	
The Research	Creating metaphors is an effective way to identify similarities and differences, a strategy shown to be effective at increasing student achievement (Marzano, Pickering, and Pollock 2001).
Classroom Implications	• You can use metaphor creation to help students identify similarities and differences, a prime strategy to improve achievement. Metaphors are effective both when you explicitly guide metaphor creation and when students independently create them. Direct instruction is more effective if there are specific similarities and differences on which you want your students to focus. On the other hand, if divergence in student thinking is your goal, have students work independently (Marzano, Pickering, and Pollock 2001). • When you direct the creation of the metaphor, provide the first element and the abstract relationship. That will scaffold the student's construction of the metaphor. Once students become familiar with these abstract relationships, you can give students the first portion of the metaphor and let them develop the second component as well as identify the relationship (Marzano, Pickering, and Pollock 2001). • Because the relationship in a metaphor is abstract, it is important that the instructional strategies you use that involve metaphors address this abstract relationship (Marzano, Pickering, and Pollock 2001).
Application Example	When studying how organisms get the energy they need to live, use various metaphors to develop understanding. Examples might be comparing an organism to a factory or to a car. Use multiple examples with your students and have them identify the strengths and weaknesses for each set of comparisons.
Technology Applications	NA
Resources	• *Understanding Models in Earth and Space Science* (Gilbert and Ireton 2003) • *Metaphors and Analogies: Power Tools for Teaching Any Subject* (Wormeli 2009); online at *http://pwoessner.com/2009/12/17/metaphors-and-analogies-power-tools-for-teaching-any-subject*

(continued)

Instructional Tool 2.7 *(continued)*

5. Visual Models: Graphs, Pictures, and Diagrams	
Visual models include images of actual objects (photographs) or graphics of objects, graphs, and or other two-dimensional representations of ideas or data. Different visual models are more or less effective for different purposes and yield different understandings (e.g., photographs to capture images and graphs to display relationships).	
The Research	Graphic models can be used to determine students' preconceptions. Presenting a graph, table, or figure and asking students to describe/interpret the image allows you to determine what they know and/or misunderstand about the represented ideas. This can be done individually or in small groups (Wright and Bilica 2007). If you have your students share and critique their developed models, it will help them see connections between ideas and help you recognize gaps and changes in student understanding (Windschitl 2008).
Classroom Implications	Give students multiple opportunities to develop figures, graphs, and charts during experimentation because they are part of, and help explicate, students' mental models. When students share their models in small groups, alternative explanations will be heard, allowing students to confront their preconceptions and those of their peers.
Application Example	For almost any topic, you can display preinstructional graphic models and ask your students to explain them in their science notebooks. As the lesson progresses, have them re-visit the models and explain again. You can also show a picture (e.g., a ball on a ramp) and have students make predictions about what will happen if the ball gets a slight push and explain their reasoning.
Technology Applications	Free graphing software is available, including *http://nces.ed.gov/nceskids/createagraph*. The Computations Science Education Reference Desk (*www.shodor.org/refdesk*) is a resource for free tools for creating graphs, calculators that plot changes between dependent variables, lots of simulations and computational models, and software for creating computational models.
Resources	• *Stepping Up to Science and Math: Exploring the Natural Connections* (Goldston 2004) (also available from NSTA as an e-book) • *Activities Linking Science with Math, K–4* (Eichinger 2009a) and *Activities Linking Science with Math, 5–8* (Eichinger 2009b) • *Developing Visual Literacy in Science, K–8* (Vasquez, Comer, and Troutman 2010)

Instructional Tool 2.7 *(continued)*

6. Dynamic Models	
\multicolumn{2}{Dynamic models are visualization and analysis tools that help students detect patterns and understand data. Examples are simulations, computer-based models, geographical information systems, and animations.}	
The Research	Computer simulations can enhance students' conceptual understandings, as well as improve achievement with complex concepts, more quickly than traditional instruction (Cawelti 1999). They are helpful when instruction involves scientific models that are difficult or impossible to observe and can simplify complex systems. Simulations are most effective when used by students individually or in small groups, resulting in better conceptual understanding. Use of simulations also appears to increase problem-solving and process skills (Cawelti 1999).
Classroom Implications	• You can use simulations to probe students' preinstructional ideas (Hand 2006). Simulations can be used in whole-class, small-group, and individual instruction, though they are most effective in small groups or when used by individual students (Cawelti 1999). But be careful. Use of simulations can promote misconceptions unless you explicitly work with your students to identify limitations of the simulated model (Cawelti 1999). • If you lack expensive laboratory equipment, use virtual manipulatives (Hartman and Glasgow 2002).
Application Example	If you complete a unit on food and nutrition and want your students to further explore the ideas of calories and weight, you can use the "Eating & Exercise" simulation available for free from the Interactive Simulations website at the University of Colorado at Boulder. Students can determine the number of calories in their favorite foods, the amount of exercise required to burn off calories, and the relationship between calories and weight. This simulation can be run online or downloaded at *http://phet.colorado.edu/en/simulation/eating-and-exercise*.
Technology Applications	Model-It, developed at the University of Michigan, is designed specifically to make systems diagramming and modeling software accessible to precollege students. Information and research about modeling in general and Model-It in particular are available at *www.umich.edu/~hiceweb/modelit/index.html*.
Resources	Some virtual manipulative sites are • Wonderville at *http://support.wonderville.ca/v1/home.html* • Utah Education Network at *www.uen.org/3-6interactives/science.shtml* • Global Classroom's collection at *www.globalclassroom.org/ecell00/javamath.html*

(continued)

CHAPTER 2: Implementation of the Framework

Instructional Tool 2.8

Sense-Making Approaches: Nonlinguistic Representations—Visual Tools

Brainstorming Webs	
\multicolumn{2}{Brainstorming webs are open systems that help students think "outside the box," creating webs related to their own personal thinking, yet also allowing them to move from idea generation to organization and transformation. Brainstorming is used at the beginning of a process but webs can be revised as thinking changes (Hyerle 2000). Webs are holistic and usually unstructured. They start with a central idea and support free association to create a graphic that reflects relations with other ideas (Young n.d.). Types of webs include clustering, mind mapping, and circle maps.}	
The Research	• Clustering as a pre-writing strategy builds strong links among associative thinking, drawing, creativity, and fluency of thinking (Rico 2000). Mind mapping connects brain hemispheres, drawing on creativity and logic in the development and support of memory and depth of understanding. • Mapping draws from students' prior knowledge, which is essential for student transfer of information to new contexts, and allows them to connect new information to their maps (Hyerle 2000). • Circle maps purposefully make no connections, leaving the brain free to brainstorm and later make connections (Hyerle 2000). Generating group circle maps enhances accountability and shared ownership (Lipton and Wellman 1998).
Classroom Implications	• *Clustering* uses ovals and words to generate ideas, images, and feelings around a stimulus word. Begin with an idea, write a phrase or word in the central oval, branch out to other ovals and add words, and extend these ovals by adding details or new ideas. Clustering requires no drawing abilities and is a good starting point to develop mental fluency. Teachers may use clustering as a whole-class, small-group, or individual activity. Clustering is not a structured organization of ideas but rather a network of associations. Initial clusters can serve as a foundation for the development of more-focused webs that lead to greater clarity of thinking and writing. • *Mind mapping*, a more specific technique than clustering, supports creativity and memory by using both words and images to represent relationships and conceptual knowledge (thus, it depends on both sides of the brain). Students start in the middle of the page with a word or drawing for a concept, write on arched lines to build connections between ideas, and draw connections among parts of the map. This strategy determines prior knowledge but can also be used as a lesson progresses. It is useful when studying content-area textbooks, as students can take notes that show both the big picture and the details. It also shows interrelationships among concepts over the course of a text (Hyerle 2000).

Instructional Tool 2.8 *(continued)*

	• *Circle maps* help students focus on a topic and brainstorm related ideas while framing them in context, thereby helping students understand their own and other students' points of view. Circle maps are best used in small groups using chart paper (Lipton and Wellman 1998). They consist of a circle within a circle on a sheet of paper. Students place a word, symbol, or picture that represents the concept or idea being studied in the center circle. In the outer circle, students list words and phrases that relate to that concept or idea. Students write, outside the circle, information about their lives that provides the context for their ideas (Hyerle 2000). They list in the upper right-hand corner categories of ideas generated so far. In the lower left-hand corner they list frames of reference. To help students list frames of reference, teachers ask questions such as, "What types of things influence your point of view—for example, prior knowledge or personal and cultural influences?" (Lipton and Wellman 1998).
Application Example	Student groups begin with the word *force* in the center of a circle map. They brainstorm, adding words, phrases, and/or pictures in the outer circle. They list categories in the upper right-hand corner for the words they listed in the outer circle. This gives you an idea of their prior knowledge and helps launch the lesson, based on their current understandings.
Technology Applications	Software is available for each of these strategies (see Resources below). Content Clips from NSDL is a website that allows import of photos and video and provides tools to sort and organize conceptually. Regardless of software used, Smart Boards are always an option for sharing and interacting.
Resources	• Rapid Fire in Inspiration (*www.inspiration.com*) • Content Clips from NSDL (*www.contentclips.com*) • Clustering software (*http://bonsai.hgc.jp/~mdehoon/software/cluster/software.htm*) • iMindMap (*www.imindmap.com*) • FreeMind (*http://freemind.sourceforge.net/wiki/index.php/Main_Page*) • Thinking Maps website (*www.thinkingmaps.com/htthinkmap.php3*)

(continued)

Instructional Tool 2.8 *(continued)*

Task-Specific Graphic Organizers	
	Task-specific organizers help students see the big picture about the content you want them to learn. They help organize the mind and promote "thinking inside the box" (Hyerle 2000). They are applied in formal, rule-based ways and used for defined tasks or in a specific knowledge area. Emphasis is on organization as specified by the teacher rather than the learner's creative organization (Young n.d.). Organizers can be descriptive, sequential, process or cause and effect, categorical, comparison or relational, and problem solution.
The Research	• These highly structured tools help develop habits of mind such as persistence, self-control, and accuracy and precision of language and thinking. Their step-by-step nature results in concrete models, thereby providing scaffolds for students who might otherwise give up. They offer a global view of a process as well as an end-point and help students stay "inside the box" of learning that is the target. They also result in a written display of students' ideas, which students can reflect on and perhaps use to modify their thinking. Once students become familiar with these structured organizers, they are more able to create their own organizers and control their own thinking (Hyerle 2000). • Flowcharts, a type of sequential organizer, can represent simple one-way processes (e.g., a simple experiment), more complex scientific processes with loops and decision points, and cause and effect (Gore 2004). • Venn-Euler diagrams help students learn to compare and contrast because the abstract is made visible, supporting reasoning (Gore 2004).
Classroom Implications	• *Descriptive organizers* describe persons, places, events, and things. They have a central, main idea with subcategories or properties radiating out from the center (Marzano, Pickering, and Pollock 2001; Gregory and Hammerman 2008). They are simple visual representations of key concepts and related terms and ideas; they let students see relationships among ideas and how the ideas link together. They also help students represent abstract ideas, show relationships, and organize ideas for storage and recall. • *Sequential organizers* organize events in a sequence and include flowcharts, timelines, and cycle diagrams. Flowcharts are useful when teaching a process with several steps. They can represent simple one-way processes (e.g., a simple experiment), more complex scientific processes, or cause and effect (Gore 2004). • *Process or cause-and-effect organizers* show either cause-and-effect relationships or a sequence of causal events. They describe how events affect one another in a process. Students identify and analyze the cause(s) and effect(s) of an event or process. There are many different types of cause-and-effect organizers so it is important to select the one that best fits the content you are teaching.

Instructional Tool 2.8 *(continued)*

	• *Categorical organizers* are used for classification and have a horizontal or vertical treelike organization, showing a system of things ranked one above another or left to right. They can be used at the beginning of a project to visually arrange interrelated and sequentially ordered sections within a whole. Projects, term papers, and study of systems all work well with hierarchical organizers. They are used to show causal information, hierarchies, or branching procedures. • *Comparison or relational organizers* identify similarities and differences or comparisons among objects or events. Overlapping Venn-Euler diagrams are used to compare and contrast (when they overlap). They help develop logic, deductive reasoning, and cognitive processes by having students look at various relationships among classes (i.e., a set, a collection, or a group of words or concepts) such as circles within circles and circles outside of circles. Matrixes help when two or more things are compared and contrasted (Gore 2004). • *Problem-solution organizers* identify a problem and possible solutions. They show the problem-solving process by defining the parts of the problem and possible solutions. They structure a process to identify a problem, identify a goal and ways to perceive the goal, identify constraints and effects on the problem context, and generate solutions and text alternatives. The organizer "gives a flow of possible solutions and pathways back when a solution is not immediately apparent" (Hyerle 2000, pp. 71, 74).
Application Example	Students use a descriptive organizer to make sense of their initial ideas about states of matter. "States of matter" is written in the center of their papers. Students then web out ideas, including definitions and interactions. The web is modified during the course of study as students learn more.
Technology Applications	Regardless of software and/or strategy used, Smart Boards are always an option for sharing and interacting.

Resources	• Find organizers at
	   • *www.graphic.org/goindex.html*
	   • *www.sdcoe.k12.ca.us/score/actbank/torganiz.htm*
	   • *edhelper.com/teachers/Sequencing_graphic_organizers.htm*
	   • *www.educationoasis.com*
	   • *www.enchantedlearning.com/graphicorganizers*
	• Find graphic organizer generators at *www.teach-nology.com/web_tools/graphic_org*.
	• A good resource on all types of graphic organizers is *A Field Guide to Using Visual Tools* (Hyerle 2000).

(continued)

CHAPTER 2: Implementation of the Framework

Instructional Tool 2.8 *(continued)*

Thinking-Process Maps

Thinking-process patterns grow out of and synthesize brainstorming webs and graphic organizers, and they support "thinking about the box." They help students define specific thinking processes as recurring patterns that can be transferred across disciplines, and they guide the building of simple and complex mental models. Students focus on evaluating their own and peers' mental models and reflect on their own meaning making (Hyerle 2000). The term *thinking-process maps* refers to concept maps, systems diagrams, and thinking maps.

The Research	• Concept maps show relationships between concepts; students derive meaning from seeing these relationships (Novak 1996). Concept mapping helps people learn how to learn, differentiate misconceptions from accurate conceptions, decrease anxiety, and improve self-confidence (Fisher, Wandersee, and Moody 2000; Hartman and Glasgow 2002). It has positive effects on student attitudes and achievement (Horton et. al 1993; Cawelti 1999). It promotes metacognition, especially when used with interviews, as it provides a frame of reference for students to analyze their own thinking, identify their strengths and weaknesses, and set learning targets. Mapping also increases student motivation (Stow 1997).
	• Students in classes involved with group concept mapping outperform students in classes where maps are created individually or not at all (Brown 2003). There appears to be no difference in achievement when the maps are made by teachers or by students. However, greater achievement is demonstrated if students supply the key words in concept map construction (Cawelti 1999). Concept maps are especially effective when working with concept-rich units (Brown 2003). They are also useful assessments because they help determine changes in understandings of concepts and connections among them (Cawelti 1999).
	• For most people, it is easier to interpret and remember images than text, and drawing a diagram shows the linkages among concepts or variables better than text. Connecting new information to existing knowledge by using diagrams helps stimulate thinking about the situation. Diagramming can also overcome language barriers. Constructing diagrams as a group aids brainstorming, analysis, communication, and understanding (ICRA n.d.; Vasquez, Comer, and Troutman 2010).
	• Thinking maps focus on forms of concept development and reflection, so it is important for teachers to facilitate the development of four habits of mind: questioning, multisensory learning, metacognition, and empathic listening (Hyerle 2000).

Instructional Tool 2.8 *(continued)*

Classroom Implications	• *Concept maps* are useful at any point during a unit of study and are effective evaluation tools because they require high levels of synthesis and evaluation (Novak 1996). You can make a concept map, have students make them individually, or ask small groups to develop them (Cawelti 1999). You should model concept mapping for students. Then use cooperative learning to let students model the techniques for others (Hartman and Glasgow 2002). Group construction of concept maps makes evident students' misunderstandings, allows them to correct one another's mistakes, and develops deeper understanding (Brown 2003). Have students work individually on maps and then work collectively to merge their maps into a more comprehensive group map. This process provokes dialogue and debate and keeps students on task during collaborative efforts (Novak 1996). Questioning enhances the effectiveness of concept maps. A good focus question leads to a richer map; questions you ask during map construction should probe student thinking and guide instruction (Cañas and Novak 2006). • *Systems diagrams* are used to represent ideas about complex situations and help us make sense of the world. They are used to describe either a structure or process, but not both. Systems diagrams have many forms and uses, but when studying a system they can be considered a "model" (Mind Tools n.d.). You can use them for brainstorming, but they are also helpful as students try to understand connectivity in a system. They can also be used to diagnose, plan and implement, and communicate (Mind Tools n.d.). Simple diagrams with 5 to 10 elements are best, though it is difficult to limit a diagram to these few elements. Include only essential elements and use single words and short phrases. A variety of systems diagrams are available with different purposes (see Resources on next page). • *Thinking maps* are often highly structured and thus resemble task-specific organizers. However, they are different in that they help students see the big picture because students must analytically organize material. Thinking maps are, in many ways, a synthesis of brainstorming webs and graphic organizers. The eight different kinds of thinking maps—circle maps (define context), tree maps (classify/group), bubble maps (describe with adjectives), double bubble maps (compare/contrast), flow maps (sequence and order), multi-flow maps (analyze cause and effect), brace maps (identify part/whole relationships), and bridge maps (draw analogies)—focus students' thinking on the map as well as on what influences the creation of the map. Implementation requires use of recurrent thinking patterns and reflective questioning (Hyerle 2000). The maps can be used individually or in concert.

(continued)

CHAPTER 2: Implementation of the Framework

Instructional Tool 2.8 *(continued)*

Application Example	Provide students with 10 to 20 concepts to map for a given topic of study. Ask them to determine which concepts are the most significant (superordinate concepts) and also identify the subordinate concepts and appropriate linking words to describe the concept relationships. Give them multiple opportunities to synthesize and evaluate their maps. Then ask them to add to their maps several more related concepts. These steps require them to recall, synthesize, and evaluate (Novak 1998).
Technology Applications	• Inspiration mapping software can be used in a variety of ways. Information can be found at the Inspiration Software, Inc. website: *www.inspiration.com*. • Content Clips is an NSDL project that provides for import of photos and video and then offers tools for sorting and conceptual organizing (*www.contentclips.com*). • Model It, developed at the University of Michigan, is designed specifically to make systems diagramming and modeling software accessible to precollege students. Information and research about modeling and Model It are available at *www.umich.edu/~hiceweb/modelit/index.html*. • There are free online versions of cognitive mapping tools such as *bubbl.us* (*http://bubbl.us*) and Mind Meister (*www.mindmeister.com*). • ThinkingMaps software is also available (see Resources below). Regardless of software used, Smart Boards are always an option for sharing and interacting.
Resources	• See *Learning, Creating, and Using Knowledge: Concept Maps as Facilitative Tools in Schools and Corporations* (Novak 1998). • Visit *www.socialresearchmethods.net/mapping/mapping.htm* for a concept-mapping resource guide. • Go to *http://systems.open.ac.uk/materials/t552/index.htm* for an excellent tutorial that will help you learn about various systems diagrams and their uses and construction. • A full open-learning tutorial on systems diagramming is available at *http://openlearn.open.ac.uk/course/view.php?id=1290&topic=all*. • *A Field Guide to Using Visual Tools* by David Hyerle (2000) includes a full chapter (Chapter 6) on thinking maps. • Go to *www.thinkingmaps.com/htthinkmap.php3* for information about the eight thinking maps developed by David Hyerle (2000): circle maps for defining context, tree maps for classifying and groups, bubble maps for describing with adjectives, double bubble maps for comparing and contrasting, flow maps for sequencing and ordering, multi-flow maps for analyzing causes and effects, brace maps for identifying part-to-whole relationships, and bridge maps for seeing analogies. • See a video introduction to thinking maps by Dr. Pat Wolfe, together with examples of maps and podcasts, to better understand how to use thinking maps: *www.opencourtresources.com/thinking_maps*.

Instructional Tool 2.9

Sense-Making Approaches: Nonlinguistic Representations—Drawing Out Thinking

Drawings and Annotated Drawings	
The act of drawing uses the right brain to visualize and solve problems and allows thinking in a visual language. Students' drawings can be used by teachers to assess science concept knowledge, observational skills, and ability to reason. Drawings also allow students to explore their own understanding about a concept.	
The Research	• Our brains receive 80–90% of information visually. When drawing, students are free to express much of this information, enhancing the nonlinguistic representations in our students' minds as well as their content understandings (Hyerle 2000; Marzano, Pickering, and Pollock 2001). • Drawings are considered more fair (less biased) than more structured approaches to expression because students can choose what they draw and because drawings are related to students' own experiences (McNair and Stein 2001). • Annotated drawings are an alternative form of expression that allows those of our students who may understand a concept, but find it difficult to express themselves in words, to convey what they know. They also allow students to reveal understandings that might not otherwise be revealed (Atkinson and Bannister 1998). • We can use drawings to uncover how our students perceive objects and the degree to which they perceive and represent details. In addition, the drawing process elicits student questions on points that their peers and the teacher can then clarify (McNair and Stein 2001). • Drawings are a good way to promote conceptual change because they provide information about specific misconceptions, help our students grapple with their own ideas and questions, and provide information that shows development of ideas over time (Edens and Potter 2003; Stein and McNair 2002). • Drawing activities, coupled with interviews, help teachers explore students' ideas about abstract concepts (Köse 2008). Drawing tasks and annotated illustrations support the processes of selection, organization, and integration—all cognitive processes necessary for meaningful learning. But keep in mind that the effectiveness of drawings may depend on the level of a student's prior understandings (Edens and Potter 2003).

(continued)

Instructional Tool 2.9 *(continued)*

Classroom Implications	• Teachers typically present mental models and scientific thinking to their students using verbal language, but drawings are a viable alternative to explore their thinking and should be used more often. A combination of visual and verbal strategies may be best used when teaching nonobservable science concepts such as energy. This combination lets us look for a match between our students' verbal and visual representations and lets our students elaborate on their understandings more fully than if only one strategy were being used (Vasquez, Comer, and Troutman 2010). • Students can discuss what they have drawn and may, thereby, reveal their misconceptions (Edens and Potter 2003). They can later revisit their drawings, reconstruct earlier concepts, and use drawings to rethink an idea (McNair and Stein 2001). • Student-created drawings provide us with information that helps determine activities that will best serve students' learning needs (Stein and McNair 2002). Because drawings are based on students' own experiences, they help teachers to be responsive to students' interests, background knowledge, and emerging skills (McNair and Stein 2001).
Application Examples	"Talking drawings" translate mental images into simple drawings (McConnell 1993). Students are asked to (1) create mental images of their understandings of a topic prior to instruction, (2) draw pictures representing their mental images and label them, (3) after content instruction, draw a second round of pictures and label them, and (4) write about how their drawings have changed. This strategy makes student construction of knowledge visible, allows students to check their own understandings, and lets them adjust their thinking and study habits (Scott and Weishaar 2008).
Technology Applications	Any graphics program or photo-editing program can be used. There are many free ones.
Resources	*Science Formative Assessment: 75 Practical Strategies for Linking Assessment, Instruction, and Learning* (Keeley 2008)

Instructional Tool 2.9 *(continued)*

	Concept Cartoons
	Concept cartoons are cartoon-style drawings that present alternative conceptions in science, elicit students' ideas, and challenge their thinking to promote further development of their ideas.
The Research	• Concept cartoons (see example on p. 69) work in a variety of teaching situations and across grade levels. They call on students to focus on constructing explanations for the different situations in the drawings. Students must choose between the different explanations in the cartoon, either individually or in small groups. This process makes it evident to them the need for investigation to answer the questions. They become responsible for choosing what is appropriate to investigate, lessening our need as teachers to respond to each student individually about their ideas (Keogh and Naylor 1999). • Concept cartoons make it possible to elicit our students' ideas either concurrently or consecutively with the restructuring of their thinking, making the learning process more continuous (Keogh and Naylor 1999). The cartoons give our students the opportunity to discuss the causes of their misconceptions, create an environment where they can all participate during class discussion, and activate them to support their ideas (thereby remedying their misconceptions) (Ekici, Ekici, and Aydin 2007).
Classroom Implications	Concept cartoons are highly motivating because they present cognitive conflict through the alternative explanations shown in the cartoon (Keogh and Naylor 1999). Such cognitive conflict challenges students' misconceptions (an important part of our framework). It also provides a wonderful entry into inquiry experiences because students have to consider what they might investigate to clarify their understandings and determine which explanation in the cartoon is most correct. The authors encourage you to study the various concept cartoons that align with the various standards you address when teaching and use them as entries into inquiry. You can also create your own concept cartoons quite easily once you are familiar with the common misconceptions about a topic.

(continued)

CHAPTER 2: Implementation of the Framework

Instructional Tool 2.9 *(continued)*

Application Examples	*Concept Cartoon: Where Does a Plant's Mass Come From?* **Question:** This large tree started as a little seed. What provided most of the mass that made the tree grow so large? Speech bubbles: "I think most of it came from nutrients in the soil that are taken up by the plant's roots." / "I think most of it came from the Sun's energy." / "I think most of it came from molecules in the air that came in through holes in the plant's leaves." / "I think most of it came from the water taken up directly by the plant's roots." This concept cartoon is from *Hard-to-Teach Biology Concepts* (Koba 2009, p. 131). It was used to determine students' preconceptions about photosynthesis and as an assessment tool during and after instruction to determine how students' ideas had changed.
Technology Applications	Consider using ComicLife (*http://plasq.com/comiclife-win*) or Comic Creator (*www.readwritethink.org/materials/comic*) to create your own concept cartoons.
Resources	• Visit the Concept Cartoons website at *www.conceptcartoons.com*. A rich set of concept cartoons on evolution are available at *www.biologylessons.sdsu.edu/cartoons/concepts.html*. • See also *Concept Cartoons in Science Education* (Naylor and Keogh 2000a), *Concept Cartoons in Science Education* (CD) (Naylor and Keogh 2000b), and *Science Formative Assessment: 75 Practical Strategies for Linking Assessment, Instruction, and Learning* (Keeley 2008).

Instructional Tool 2.10

Sense-Making Approaches: Nonlinguistic Representations—Kinesthetic Strategies

Hands-on Experiments and Activities and Manipulatives	
\multicolumn{2}{l}{This category involves physical movement during a science learning experience. Examples include classroom experimentation that makes use of equipment such as probeware and requires movement from one part of the lab to another, use of other manipulatives (e.g., physical models), excursions into the field, and projects that require construction of materials.}
The Research	Kinesthetic activity involves physical movement; movement associated with specific knowledge builds a mental image of that knowledge in the learner's mind (Marzano, Pickering, and Pollock 2001). Simply moving activates the brain, and if you add the requirement of communication to this action much more of the brain is involved in the learning experience (Lazear 1991). Concrete experiences engage more of the senses and activate multiple pathways to store and recall information (Wolfe 2001). Hands-on activities engage students, who must interact with materials and peers (Wolfe 2001).
Classroom Implications	Almost any concept or idea can be transformed to involve physical movement. Consider each lesson and determine ways in which experimentation and other hands-on activities can replace lecture and demonstration approaches. When students present research results, have them move in front of the class. Their movements can include demonstrations of techniques used and of samples gathered.
Application Examples	Begin a class period by modeling the use of sampling materials (e.g., collection nets, sampling containers, dissolved oxygen probes) that will be used in fieldwork outside the classroom. Students practice using the materials and then go out into the field to carry out the sampling. On their return to the classroom, they sort and identify samples. They determine the best way to present their materials during a poster presentation, prepare the posters, and present the posters in a round-robin sharing session. Notice that each step in this series of activities involves students in movement and manipulation.
Technology Applications	Using probeware to collect data requires movement on the part of students.
Resources	Lesson Plans, Inc. (*www.lessonplansinc.com*) provides a variety of activities that make sure to address kinesthetic learners.

(continued)

Instructional Tool 2.10 *(continued)*

	Physical Movement
	These kinesthetic activities include the use of gestures, hand signals, and arm motions, as well as acting and role-playing. These are not science-specific actions, but work across the curriculum to improve acquisition and retention of information.
The Research	Kinesthetic activity involves physical movement; movement associated with specific knowledge builds a mental image of that knowledge in the learner's mind (Marzano, Pickering, and Pollock 2001). Simply moving activates the brain, and if you add the requirement of communication to moving, even more of the brain is involved in the learning experience (Lazear 1991). Physical simulations and role-playing call for physical activity and the arousal of emotions, helping in the acquisition and retention of knowledge (Wolfe 2001). Changing locations during a lesson enhances acquisition of information and memory (Jensen 1998). Gestures associated with learning new information help students retain information (Jensen 1998).
Classroom Implications	Learning almost any concept or idea can be made more engaging by requiring physical movement. Simple learning approaches such as the jigsaw technique require students to move around the room. Learning stations are another way to ensure movement. Role-playing can be used to illustrate concepts, practice skills, stimulate interest, and make ideas more concrete for discussion (Hartman and Glasgow 2002). During any lesson, students can indicate their levels of understanding by gestures (e.g., thumbs-up/thumbs-down).
Application Examples	Use four-corner synectics (Walsh and Sattes 2005) to determine students' preconceptions about almost any concept. This activity requires you to use four different metaphors for a concept (e.g., the heart is most like a/an … bucket, pump, house, or engine), placing a label for one of the metaphors in each corner of the classroom. Students take a moment to personally think about and record which metaphor best reflects their current understandings of the concept. They then go to that corner and talk with the other students at the corner about why they selected that metaphor. Students then share their thinking as a whole class. These multiple metaphors give the teacher a glimpse into the students' preconceptions and start group and whole-class discussions about the concept.
Technology Applications	Various educational games are available (see *www.supersmartgames.com*), some for Wii. A very good role-playing science game is WolfQuest (*www.wolfquest.org*).
Resources	Lesson Plans, Inc. (*www.lessonplansinc.com*) has various activities for kinesthetic learners.

Chapter 3

The Framework and Instructional Tools at the Elementary Level

CHAPTER 3: The Framework and Instructional Tools at the Elementary Level

> "Responsive or differentiated teaching means a teacher is as attuned to students' varied learning needs as to the requirements of a thoughtful and well-articulated curriculum. *Responsive teaching* suggests a teacher will make modifications in how students get access to important ideas and skills in ways that students [can] make sense of and demonstrate essential ideas and skills ... —all with an eye to supporting maximum success for each learner."
>
> —*Tomlinson and McTighe 2006, p. 18*

Chapters 1 and 2 dealt largely with two of the five natural learning systems[1]—the cognitive and reflective (or metacognitive) systems. After you have worked through those two chapters, you should be well on your way to the "thoughtful and well-articulated curriculum" referred to in the quotation above.

But teaching, as you are no doubt aware, is rarely as straightforward in the classroom as a model suggests. Sometimes students "get it" right away; sometimes they don't seem to understand the targeted ideas. In the latter case, you need to move beyond your initial instructional plan. It's in times like these that more help is called for. This chapter looks at two additional natural learning systems: emotional learning and social learning. It includes information that will help you provide multiple ways for students to express themselves and for you to continue to challenge your students (emotional learning system) as well as many examples of how to include all learners (social learning system).

Various conditions in the elementary classroom have an impact on your ability to implement the Instructional Planning Framework as described in this book. For example, you probably have various subjects in addition to science to teach in a limited time frame and you need to make the most of instructional time. Or your school district might have a very structured curriculum with an established pacing guide. Or, if your school or district uses a reform curriculum such as Full Option Science System (FOSS) or Science Technology Concepts (STC), you may already be focusing on core ideas. However, neither of these curricular approaches—a very structured curriculum or a reform curriculum—ensures that each of your students understands the concepts as you want them to. The issues discussed in this chapter support further differentiation, ideas for re-teaching, and interdisciplinary approaches you can use in conjunction with your established

CHAPTER 3: The Framework and Instructional Tools at the Elementary Level

curriculum or in lessons you develop on your own. Chapter 3 focuses on the topic of the flow of matter and energy in ecosystems that was introduced in Chapter 2.

Responding to the Needs of All Learners

Consider the quote at the beginning of the chapter, which equates differentiated teaching with *responsive* teaching. And recall that the second phase of this book's framework is the *responsive* phase. It is important to give each student the opportunity to explore phenomena related to the topic of study and make sense of those experiences—that is, you must *respond* to students' learning needs and styles.

At first glance, the Instructional Planning Framework (Figure 1.2, p. 8) may seem to be proposing that you simply teach the planned learning sequence: You identify preconceptions, elicit and address those preconceptions, provide students with sense-making experiences, and then move forward to the next learning target. It's not that simple, however. Refer to Figure 3.1. You will notice that the planned learning sequence appears linear. Indeed, if every student understands the key ideas for the first target, you can move straight on to the second target. However, if students do not understand the key ideas as

Figure 3.1

Planned Learning Sequence and Implemented Learning Sequence

Planned Learning Sequence

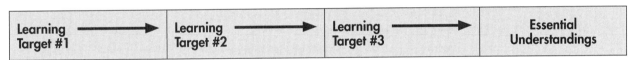

Implemented Learning Sequence

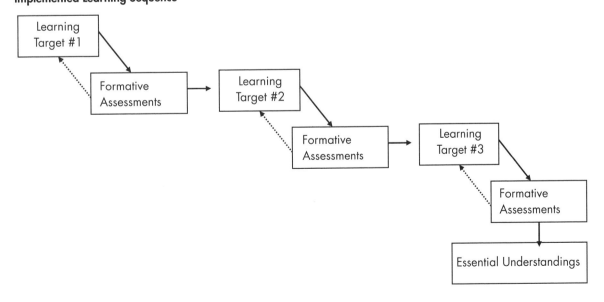

CHAPTER 3: The Framework and Instructional Tools at the Elementary Level

demonstrated by formative assessments, it is your responsibility to re-teach. This is true for each learning target until the key ideas for the topic are understood. The revisiting of key ideas makes up the implemented learning sequence.

One of the joys of elementary classroom instruction is that you probably have a single group of students you work with all year and the opportunity to get to know them well. You recognize their learning preferences and interests and their background knowledge and readiness, as well as their language strengths and challenges, and then you act accordingly. The environment you establish, your relationships with your students, and your understanding of their learning styles and needs are critical for responsive teaching.

But how can you initially plan instruction to meet the needs of each of these unique students? And what do you do if, after implementing the planned instruction for a learning target, students still struggle with the core ideas? The following two sections examine ways to plan for differentiation and look at optional formats and ideas for re-teaching. Learning Target #1 in the instructional plan for learning about the flow of matter and energy in ecosystems—that is, Food is a substance that provides fuel and building materials for organisms—is used to explore differentiation and re-teaching, but the suggestions relate to any instructional unit you teach.

Planning for Differentiation

Tomlinson (1999) recommends that teachers differentiate among three curricular elements: content, process, and product. What does differentiation look like in the plans developed in Chapter 2?

Differentiating Content

First consider the predictive phase planning template (Table 2.6, pp. 34–35). How can you differentiate this content if you want each of your students to understand the concepts? The essential understandings and learning targets must remain stable because that is the core content you are addressing (Tomlinson and McTighe 2006), but you can vary teaching methods and instructional strategies. You can also differentiate the support provided to students as they master vocabulary and skills, even though, again, the core content must remain the common learning goal. Refer to Instructional Tool 2.5 (p. 98), which addresses "reading to learn"; also in that tool, you will find research and resources for vocabulary development. Likewise, in terms of skills, after you determine what the basic level of mastery should be, group students (or partner them) according to their skill levels. Read the scenario in Figure 3.2 for a glimpse of what such differentiation looks like in a classroom that is studying the flow of matter and energy in ecosystems.

Inquiry instruction and support of metacognition also enhance differentiation. In fact, an inquiry focus and a metacognitive focus should be included in every instructional unit. Inquiry is a methodological approach central to the science classroom and the Instructional Planning Framework. It supports differentiation in several ways: It fosters critical thinking and reasoning skills and it develops students' abilities to acquire

Figure 3.2

Differentiating Content: Possible Variations for a Class Studying "The Flow of Matter and Energy in Ecosystems"

Mrs. Keenan knows that there is a wide range of abilities among her students. She also is aware that there are very complex concepts in her instructional unit on the flow of matter and energy in ecosystems. But she is determined that each student will understand these ideas, starting with the understanding that food is a substance that provides fuel and building materials for organisms (Learning Target #1).

During the card sort at the beginning of the lesson, she uses, whenever possible, index cards that include not just words but images. She finds texts at different reading levels and supplemental materials on the subject of the scientific definition of *food*, and she establishes reading partners to support those students with learning disabilities and those who are English-language learners. Mrs. Keenan also highlights essential text in the resources about organisms to assist her students with lower reading skills. She uses a teacher read-aloud for the more complex passages in the reading materials to help those of her students with auditory preferences and provides taped versions of the text for students who have attention problems or difficulty reading nonfiction.

Mrs. Keenan uses the essential vocabulary from her plan for basic students and expands her lists for more advanced groups (e.g., she adds specific names for the various organisms, such as *primary consumer* and *secondary consumer*). During vocabulary instruction, she uses strategies that help students with varied learning styles. She decides to focus in this unit on similes and analogies to help her students understand the core vocabulary. Finally, she establishes a word wall that uses words and images to help students with cognitive processing or attention problems.

---

and apply science concepts and to effectively communicate their understandings. When we give students the opportunity to ask their own questions and answer them, we have provided an avenue for differentiation (Gregory and Hammerman 2008).

Self-regulated thinking, including self-evaluation (see Instructional Tool 2.2, p. 83), is central to metacognition. Self-assessment enables you and your students to focus on the established, nonnegotiable goals for the class, as well as on individual goals for each learner (Tomlinson and McTighe 2006). Your support of students as they improve their self-regulation skills during inquiry-based instruction will enhance differentiation in your classroom.

Differentiating Process

The ways in which students access the targeted content is key to their learning (Hall, Strangman, and Meyer 2009), and you have many options to improve access. Curriculum components you can use include (1) the learning activities themselves, (2) the resources used to access the content, (3) grouping practices, (4) modifications

CHAPTER 3: The Framework and Instructional Tools at the Elementary Level

based on learner needs, and (5) time and space modifications (Tomlinson et al. 2002 ; Tomlinson and McTighe 2006). See Table 3.1, p. 134, for elements to modify for differentiation and for differentiation strategies.

1. **Learning Activities:** The first two chapters of this book focused heavily on selecting learning activities and research-identified strategies that elicit and confront students' preconceptions and help them in sense making. In particular, those chapters emphasized the importance of using activities that call for analytical, critical, and creative thinking skills (Instructional Tools 2.1 and 2.2), as well as for having students apply what they learned (Tomlinson et al. 2002). Many of the strategies in the other Instructional Tools (Tools 2.4–2.10) also vary the ways in which content is presented and, therefore, the ways in which students access that content. Coupled with a range of teaching methods (e.g., expository, demonstrations, discussion, inquiry), varied activities make it more likely that more students with diverse learning and thinking styles will be able to understand and learn the content you teach (Gregory and Hammerman 2008). Review the responsive phase plan (Table 2.15, p. 60) and notice the variety of methods and strategies employed.

2. **Resources:** You can also vary the resources used by students during the learning activities, during assessment, while developing products, or during extension activities. Students might use resources independently, but they occasionally will require your support or that of other people. Reach out to parents, experts, peers, other school personnel, university faculty, and community workers for assistance, remembering that among your *parents* you may also find experts, university faculty, and community workers. Other (nonhuman) resources include print materials (e.g., books, magazines, tables, graphs, and the web), as well as nonprint materials (e.g., models, photographs, software, equipment, video and audiotapes) (Tomlinson et al. 2002). It is important to provide multiple examples related to content, and if these examples are in multiple formats, so much the better. For instance, if you are teaching about various organisms in a food chain, you can use reading materials at various levels, static images, video, computer animations, and living organisms that were observed during fieldwork or are being kept in the classroom. Providing a wide range of resources and formats serves the needs of a diverse group of learners.

3. **Grouping Practices:** You should vary grouping occasionally—at different times, assign work for the whole class, for small groups, for partnerships, and for individuals. For instance, during instruction on the first learning target for flow of matter and energy in ecosystems in Chapter 2, the teacher showed the class a concept cartoon (Figure 2.11, p. 69). Students were asked to pick one of four answers to the question "What do you think happens to the food that animals eat?" Individual students recorded their answer choices and reasoning for the choices in their science notebooks. Small-group discussions followed. The students discussed each

group member's choice, reached consensus on a group choice, and developed a rationale for their choice. Groups then shared their choices and reasoning with the entire class. The teacher wrote down and displayed a list of questions students still had about the topic. Finally, each group chose a question to investigate.

It is important to use flexible grouping and small-group instruction on a regular basis because students benefit from working together toward a common learning goal (Marzano, Pickering, and Pollock 2001; Tomlinson et al. 2002). You should change group formats frequently based on the particular instructional goals, and on student interest, readiness, learning styles, and even resource availability (e.g., the number of books or models on hand, the availability of classroom aides). Whether you determine group composition or let students choose groups depends on the content as well as on student learning needs.

4. **Modification Based on Learner Needs:** As you are well aware, there are ranges of learning and thinking styles, as well as individual learning needs, among your students. It is worth your time, early in the school year, to have students complete style and interest inventories and use these findings to plan for instruction. (This chapter does not cover ways to inventory styles and interests, but you will find various ideas and tools for using such inventories among the resources at the end of the chapter.) Once you become aware of the various learning styles of your students, it is important to determine their content backgrounds and readiness-to-learn before introducing a particular topic. Recall the anticipation guide that was used in the flow of matter and energy unit to identify student preconceptions (Figure 2.7, p. 49). A guide such as this one provides a good understanding of students' backgrounds related to a particular content.

Style and interest inventories, as well as pre-assessments of content understandings, are starting points for varying approaches to meet the needs of your students. It is also essential that you *continuously* assess your students' understandings and adjust instruction *along the way*. You can adjust the learning activities themselves, grouping, student products, rubrics, resources, coaching or scaffolding practices, pacing, space, and extension activities (Tomlinson et al. 2002). Your constant attention to the understandings demonstrated by your students is what shifts the planned learning sequence to the implemented learning sequence. Attention to your students' needs, coupled with multiple options for them to take in information and make sense of ideas, can increase the potential of your students to learn.

5. **Time and Space Modifications:** Time requirements for students vary tremendously based on background knowledge, learning styles, and specific learning challenges. It is important to be flexible in setting time requirements if you want all students to learn the essential understandings of any science unit. Offering flexible deadlines and setting longer time periods for completion of certain tasks can help students who work more slowly or who are weak in certain skills. You can also

CHAPTER 3: The Framework and Instructional Tools at the Elementary Level

Table 3.1

Elements to Modify for Differentiation and Differentiation Strategies

Elements to Modify for Differentiation	Differentiation Strategies*	
Grouping Practices Configurations to enhance knowledge and skills acquisition	There are many methods to manage groups and partners and to easily shift composition. • Clock Buddies generate varied partnerships that are easy to shift. www.readingquest.org/strat/clock_buddies.html • Seasonal Partners is a grouping strategy in which students find a partner for each season of the year from lists provided by the teacher. When doing partner work, the teacher asks students to work with their "winter," "spring," "summer," or "fall" partners. To help in differentiation, seasonal partners can be pre-assigned based on, for example, student interest, readiness, or ELL status. You can easily shift students to a new partner (or group) based on student interest and learning inventories. • There are many other strategies to generate random groups. See www.cte.unt.edu/home/classroom/Cooperative%20Grouping%20Strategies.pdf. • Regardless of grouping choice and configuration, it is critical that you give groups clear, written directions for their work and support them during the learning experience. • Cooperative grouping guidelines (provided in writing to each group) help students work well together as well as develop targeted understandings and social skills. www.muskingum.edu/~cal/database/general/group.html	
Learner Need Modifications Variations based on students' interests, readiness to learn, ELL status, and preferred learning styles	There are many strategies to modify instruction based on student need. Descriptions of several key approaches follow, and resources are listed in the Build Your Library section of this chapter, beginning on page 154.	
	Choice Boards	• Choice boards allow the teacher to change assignments in permanent pockets. Students choose an activity from a targeted row. This lets the teacher target readiness levels yet gives students some choices. • Tic-Tac-Toe boards are similar but instead of using pockets they are developed on posters for each topic.
	Learning Contracts	Learning contracts allow students to work somewhat independently on material that is largely teacher-directed. They often allow some choice. www.saskschools.ca/curr_content/bestpractice/contract/index.html
	Curriculum Compacting	Pre-assessment data is used to identify what a student already knows on a topic, exempts the student from work on that topic if he or she has acquired the essential understandings, and plans for meaningful and challenging use of his or her time. This strategy maximizes time for the more advanced learner. www2.ed.gov/pubs/ToolsforSchools/curc.html

Table 3.1 *(continued)*

Elements to Modify for Differentiation	Differentiation Strategies*	
	Centers	Centers are designed instructional spaces for specific experiences, such as learning, relearning, or extending learning. *www.saskschools.ca/curr_content/bestpractice/centres/index.html*
	Tiered Activities	Tiered activities allow students with different learning needs to work on the same concept but at different levels of complexity, open-endedness, and/or abstractness, thus offering appropriate challenges for each student. *www.saskschools.ca/curr_content/bestpractice/tiered/index.html*
Time Flexible time allotments for work	*Anchor Activities*	Anchor activities are ongoing assignments students can work on independently when they have extra time. A poster that lists possible anchor activities reminds students of their options when they find they have extra time. *www.saskschools.ca/curr_content/bestpractice/anchor/index.html*
	Personal Agendas	These personalized lists of tasks must be completed in a specific period of time. A designated time during the day is usually assigned for agenda work, but students can also work on agendas when they have extra time.
	Targeted Homework	The teacher prepares homework options that address specific concepts in advance of a lesson and assigns the options based on student need. This extends the time students have to work on specific concepts and skills.
Space Variations in space to support learning	<ul><li>Set aside quiet spots in the room to minimize noise and visual distractions.</li><li>Designate an area for independent work for students who have to make up work.</li><li>Use various room configurations and familiarize students with them so they can help efficiently rearrange the room.</li><li>Use "science centers in a box" to more flexibly use room space.</li></ul>	

Note: See additional resources in the Build Your Library section at the end of this chapter (pp. 154–156).

exempt students from deadlines who demonstrate understanding or you can compact curriculum (see Table 3.1, Learner Need Modifications). Strategies that incorporate flexible deadlines include anchor activities, personal agendas, and targeted homework (see Table 3.1, Time). You must also consider the learning space itself, using different configurations to support various student interactions as well as whole-group instruction (see Table 3.1, Space).

For an idea of what differentiated processes might look like in the classroom, see Figure 3.3, p. 136.

CHAPTER 3: The Framework and Instructional Tools at the Elementary Level

> ### Figure 3.3
> Differentiated Processes for a Class Studying "The Flow of Matter and Energy in Ecosystems"
>
> Mr. Matthews's class finished small-group experiments that were designed by the students to answer the question "Does what an animal eats become part of its body?" It was clear from the students' presentations that one entire group still struggled with the major concepts and that several students in other groups were unclear about some of the ideas. The other students, however, were ready to move forward.
>
> Mr. Matthews assigned groups who had mastered the content to various centers—a reading center, a math center, an art center, and a center that focused on an additional experiment. He used students' "winter" seasonal groups (see Table 3.1, p. 134), which were based on interest and learning style, to assign these groups. The individual students who still lacked some clarity about the content but were on the right track were assigned independent work that addressed their particular needs. While the groups and individuals worked independently or at assigned centers, Mr. Matthews worked with the remaining small group of students who struggled with the basic concepts. A bit later, one group finished the tasks at their center, so Mr. Matthews suggested that each student choose a task from his or her personal agenda and work on it while everyone else completed their work.

In Chapter 2, an instructional plan for the responsive phase (Table 2.15) was developed. It differentiates some of the five components just described, but there is still a lot of room for differentiation. Table 3.2 summarizes differentiation in the instructional plan for Learning Target #1 and also suggests a few examples of steps you might take to further differentiate. Use only one or two of these options when you first implement the instructional plan if you do not have a lot of experience with differentiating instruction. *It is better to take small, successful steps than to tackle too much the first time you teach a lesson.*

Differentiating Product

The third curricular component you can modify is lesson products. Products are the evidence of the current level of students' understanding. A basic way to differentiate products is to assess them in multiple ways in terms of students' learning and thinking styles. For example, consider using science notebooks, interviews and dialogue, arts integration, laboratory practicals, multimedia projects, and verbal presentations. You might require a portfolio compilation that includes evidence of learning that relates to the various intelligences; it might include annotated drawings, graphing, or journal reflections. Even though students have strengths in particular intelligences, they should work to improve in other intelligences (Gardner 1983, 2004). Some degree of choice could be allowed in regard to the contents of the portfolio.

Table 3.2

Differentiation in the Instructional Plan for Learning Target #1 (see Table 2.15, pp. 60–66)

	Differentiation in the Instructional Plan for Learning Target #1	**Possible Further Differentiation**
Learning Activities	• Common activities representing studied phenomena for the learning target were selected and then modified by using research-based strategies (from the Instructional Tools, pp. 42–44 and 76–126) to better engage students in the learning process and probe their thinking. • Critical thinking is supported (e.g., students must provide a rationale for responses to questions and evidence for claims during inquiry). • Inquiry is infused in the lesson, giving students opportunities to draw on their learning strengths and providing choice in experimental design.	You can tier the inquiry experience in this lesson by grouping students by readiness and providing more structured inquiry for some groups and more open-ended inquiry for others.
Resources	• A variety of resources are used to learn targeted concept (e.g., index cards, handouts with explanations of the scientific use of the word *food*, concept cartoons, or concrete materials for investigation). • Students can choose from among a variety of organisms. • Students can choose from among various measuring devices.	• Index cards for the card sorts can include images instead of (or in addition to) words to provide support to English-language learners. • Teaching aides, if available, can support students with special needs. • DVD, CDs, and audiotapes can be used in conjunction with handouts. • Simulations can be used to develop skills (e.g., Gizmo* for graphing skills).
Grouping	Various groupings (individual, partnered work, small group, and whole class) are used at various times during the lesson.	Consider learner needs when determining group and partner compositions.

* Gizmos are for-purchase, interactive simulations for math and science (grades 3–12). A 30-day free trial is available. Go to *www.explorelearning.com*.

(continued)

CHAPTER 3: The Framework and Instructional Tools at the Elementary Level

Table 3.2 *(continued)*

	Differentiation in the Instructional Plan for Learning Target #1	**Possible Further Differentiation**
Learner Needs	• An anticipation guide (see Figure 2.7, p. 49) is used to determine students' preinstructional concepts. These are used to plan the lesson. • Student conceptions are elicited early in the lesson and thinking is made evident during the lesson through class discussion, teacher questioning, and science notebook entries. • A variety of intelligences (Gardner 1983, 2004) are addressed through use of, for example, discussion, writing, concept cartoons, concept mapping and Gallery Walk, concrete materials, and experimentation.	The lesson, while differentiated to some degree, still requires students to work together in particular grouping configurations (i.e., individual, partner, small group, or whole class) at particular times. Students work through the lesson at a pace predetermined by the teacher. Table 3.1 strategies that could be infused into the lesson include • choice boards for summative assessment that aligns with learning style or learning intelligence. • tiered assignments for groups as they design experiments, providing more structure for some groups and flexibility for others.
Time	No adjustments are made in the plan.	Prepare a poster with anchor activities related to this content. Mount the poster on a wall in the classroom. When individuals or groups finish work ahead of others, they can turn to these activities.
Space	Though various grouping configurations are used, no adjustments for use of space are included in the plan.	Work with students to quickly shift furniture from whole-class to small-group configurations and laboratory setups.

The criteria to demonstrate understanding found in the predictive phase planning template (see Table 2.6, pp. 34–35) must remain the same for all students. But you can differentiate the more formal formative and summative assessments you develop to align with these criteria. Some options to differentiate assessment include negotiated assessment, tiered product assignments and tasks, choice of products or tasks by students, and portfolios (Table 3.3).

All students can acquire the core understandings of an instructional unit, but they may travel different routes to get there—with paths of varying complexity, open-endedness, and abstractness. The central elements remain the same, but the demonstration of understanding varies. In general, multiple and varied assessments meet the needs of the diverse learners in your classroom.

Table 3.3

Ways to Differentiate Assessment

Mode of Differentiation	General Description	Specific Suggestions
Negotiated Assessment	Negotiated assessment supports student-centered learning by involving students at two stages: (1) *before the assessment,* choosing the assessment task (students choose from options determined by the teacher) or setting the task (students have input in determining the options), discussing criteria, and setting criteria; and (2) *after the task is completed* (self-assessment comments; peer-assessment feedback; suggesting, negotiating, and assigning self- and peer-assessment scores (O'Neill and McMahon 2005).	If you consider using negotiated assessment, it is a good idea to start small. For example: • Involve students in setting criteria for a particular product or task. • Focus on student self-assessments and/or peer assessments. • Use a feedback form similar to Figure 3.4 on next page.
Tiered Product Assignments and Tasks	Tiering allows students with different learning needs to work on the same concepts or processes but at different levels of complexity, open-endedness, and/or abstractness. In any single lesson, you can vary the degree of assistance, structure, or abstractness; the complexity of resources used and the product itself; the required background knowledge and skills; or the number of resources given to a learner.	During any inquiry you can do the following: • Provide a choice of questions for some students and free choice for other students. • Provide a greater choice of materials for some students but a more limited choice to guide the inquiry of other students. • Provide formats for tables and charts for the students who might need them. • Include writing frames that provide different levels of scaffolding. (*www.sciencenotebooks.org/classroomTools/templates.php*) • Limit the number of resources used by some students as they work on their explanations. • Highlight significant content in the textual materials for students who need that guidance. • Vary the degree of abstractness for the actual assessment product.

(continued)

Table 3.3 *(continued)*

Mode of Differentiation	General Description	Specific Suggestions
Choice in Products or Tasks	Choice in assessment products or tasks allows you to structure the options for product and performance while allowing students some choice. Each choice is based on a common set of criteria. Additional criteria for particular students can be separately established (see Negotiated Assessment, the first item in this table.)	Tic-Tac-Toe posters and choice boards work well. See Figure 3.5 for an example of a choice board. You can set up these posters or boards in any way you choose. For instance, students can be given choices in terms of rows, columns, or diagonals OR you can give them the choice of a row plus an extra choice from either of the other two rows.
Portfolios	Portfolios are collections of student work that show students' understandings and abilities. They motivate students because they allow choice and support ongoing assessment, as the goal is to continually improve the product.	Portfolios can focus on readiness, learning and thinking profiles, and interest. They also depend on self-reflection and are modified to reflect growth in understanding. Portfolios strongly support differentiation.

Figure 3.4

Feedback Form for Negotiated Assessments

Criterion #1

Self-Evaluation

Peer Evaluation

Teacher Comments

Criterion #2

Self-Evaluation

Peer Evaluation

Teacher Comments

Criterion #3

Self-Evaluation

Peer Evaluation

Teacher Comments

Figure 3.5

Choice Board for "The Flow of Matter and Energy in Ecosystems," Learning Target #1

Make an illustrated book that tells a story explaining the importance of food to a particular animal.	Create and play a song that provides a thorough explanation of the importance of food. Be certain to provide evidence.	Conduct an experiment that demonstrates what happens to food in an organism and share your findings with your class.
Create and record an advertisement for radio that explains the importance of food and includes evidence.	Wild Card—Your choice	Write and perform a play that shows the importance of food to different organisms portrayed in the play.
Conduct a survey about the importance of food. Summarize your findings. Compare them to what scientists say.	Interview a nutritionist and use the information you gain to make a brochure about the importance of food.	Research an organism and the food it eats. Make a presentation about the importance of the food to the organism.

Re-Teaching Options

Even if you plan for differentiation in your initial instructional plan, some of your students may not develop the targeted understandings. As you draw up the initial plan, list additional strategies to use if re-teaching is required. In the Instructional Tools on pages 42–44 and 76–126, you will find multiple strategies for eliciting and confronting preconceptions and for helping students make sense of these experiences. Simply keep some of the strategies in mind as you go forward with your initial plan—they are your Plan B or even Plan C, to be used if some students or groups need additional support to understand the concepts.

You can also consider certain formats that work well both during initial instruction and for any required re-teaching (Table 3.4, p.142). The format you choose will be more effective if it is based on data generated from interest and learning style inventories taken during the early part of the school year and, of course, on assessment data you have gathered during instruction.

CHAPTER 3: The Framework and Instructional Tools at the Elementary Level

Table 3.4

Examples of Re-Teaching Formats

Re-Teaching Formats	Description	Use in " The Flow of Matter and Energy in Ecosystems," Unit, Learning Target #1
Stations*	Use various stations around the room to differentiate instruction. You can vary the number of stations visited, the time spent there, the content addressed at the station, task completion required, and whether individuals, partners, or groups can visit the station. Rather than change all stations for every unit of instruction, you can use permanently established stations and change them only for the specific content addressed at a certain time—for example, stations for vocabulary, concept mapping, and experiments. Make sure that you include enrichment stations for students who have mastered the content.	• Set up a station for students who still struggle with the concept of food. Have them complete two of the following three activities, based on what they learned or struggled to learn in their experiments: a concept definition map, a semantic map, or a Directed Reading/Thinking Activity (DR/TA) (Barton and Jordan 2001). • Establish an art center where students create an annotated drawing that describes the importance of food to a particular organism. • Set up a station that includes various data demonstrating the increased mass of various types of animals given food. Ask students to interpret the data and find supporting evidence in their textbook or online. • Include a couple of extension stations with project options that depend on various intelligences.
Adjustable Assignments	You can adjust assignments based on assessments conducted during a lesson. They are varied by level of complexity or abstractness, readiness, learning styles, or intelligence preferences. Assign adjusted assignments to small groups or individuals. Use them not only in an initial lesson plan but also for re-teaching and extension. Use choice boards or agendas to outline options.	• Students struggling with core vocabulary and concepts are asked to develop operational definitions of the vocabulary and to work on a concept map with a buddy who has better command of the content. • Students who are approaching mastery of the content conduct additional experiments based on questions they had in their initial experiments. They summarize in their science notebooks the results of their experiments, highlighting the core vocabulary and concepts. • Students who demonstrate high mastery of the content complete a related project—for example, an article for the school newsletter about the new USDA food pyramid or an art project—in their preferred learning styles.

*Jones, D. J. 2007. The station approach: How to teach with limited resources. In *Readings in science methods, K–8: An NSTA Press journals collection*. Arlington, VA: NSTA Press.

CHAPTER 3: The Framework and Instructional Tools at the Elementary Level

Table 3.4 *(continued)*

Re-Teaching Formats	Description	Use in "The Flow of Matter and Energy in Ecosystems," Unit, Learning Target #1
Personal Agendas and Independent Study	Personal agendas (see Table 3.1) coupled with independent study address various learning needs. The teacher adds tasks to students' personal agendas that target their specific learning needs. You can differentiate both content and process in student agendas and group students who are assigned identical tasks, working with them as needed. Let students who have mastered the targeted content work on an ongoing independent study or a long-term project already listed in their agendas.	• Some students may need to work with vocabulary. Assign specific vocabulary strategies to help them master terms. • Some students may need to work with concept mapping or directed reading (see comments about stations in this table). • Assign some students one of the mysteries in the *Everyday Science Mysteries* series (see Konicek-Moran in the Build Your Library section at the end of this chapter, pp. 154–156). Mysteries related to this content can be used for re-teaching because they include embedded misconceptions. They can also be used as extension activities. • Students who have mastered the concepts can proceed to the independent studies already listed in their personal agendas.

Ties to Literacy and Numeracy (Mathematics)

Reform efforts in science, mathematics, and literacy have much in common, including an increased focus on conceptual understanding, process skills, thinking strategies, and habits of mind ("habits of mind" are dispositions employed by people to solve problems—e.g., persisting, thinking flexibly [Costa and Kalick 2000]). As a result, there is a growing overlap among the disciplines and a growing number of ways that the three disciplines can be used in support of one another. Traditional science instruction often found students reading, writing, and memorizing facts, while reform curricula that are inquiry-based expect students to learn and discover through exploration, investigation, research, and communication. Mathematics, and literacy instruction have also shifted toward requiring students to analyze and communicate.

These reforms support interdisciplinary teaching in all grades. Interdisciplinary approaches allow your students to work more like scientists, and they provide a real-world context for literacy and mathematics work. Consider using lessons that lead students to understand science content through inquiry by performing experiments, graphing results, discussing their thinking with their peers, reading text to enhance explanations, and writing explanations based on evidence. Such efforts not only improve students' abilities in reading, writing, and mathematics in the context of science but also optimize instructional time devoted to science.

CHAPTER 3: The Framework and Instructional Tools at the Elementary Level

Take a moment to read the vignette in Figure 3.6. As you read, jot down the points at which you think mathematics and literacy merge seamlessly with science instruction. Then read the next two sections, which briefly explore the relationships between science and mathematics and between science and literacy, and see if what you jotted down is reflected in those sections.

Science as Related to Mathematical Literacy and Mathematical Power

Reform efforts in mathematics have resulted in development of two terms, *mathematical literacy* and *mathematical power* (Hammerman and Musial 2008), that connect directly to the framework and approaches outlined in this book. They align with inquiry and conceptual change, which are central to the Instructional Planning Framework.

Mathematical literacy includes being able to use quantitative information and appreciate the value of mathematics. Quantitative information is essential in the science experiences you promote through inquiry. In addition, using quantitative information in the science classroom should enhance students' appreciation for mathematics.

Mathematical power means that students can use a variety of methods to solve nonroutine problems and do worthwhile work that includes student exploration, conjecture, and logical reasoning. Again, this concept resonates with and enhances inquiry-based science and development of conceptual understanding. Mathematics provides scientists (including students doing science) another system with which to share, communicate, and understand science concepts.

These reform efforts in mathematics stress mathematical reasoning, problem solving, communications, and connections to the real world. Interdisciplinary problem solving lets your students use mathematical ideas and skills in authentic ways. Science provides a unique opportunity for them to learn and demonstrate competency in both mathematics and science, because science provides a context for doing mathematics in a real-world setting (Hammerman and Musial 2008).

As a result, science instruction can concurrently build many of the mathematics understandings outlined in *Principles and Standards for School Mathematics* (National Council of Teachers of Mathematics [NCTM] 2000). The authors reviewed the NCTM standards and identified those standards with the strongest connection to science (Table 3.5, p. 146).

We focus on the 2000 NCTM standards in this chapter. However, it should be noted that new math standards—*Common Core State Standards for Mathematics* (2010)—were in the draft stage and under consideration by the 50 states at the time of this book's publication. See Table 3.6, p. 149, for selected math standards from that draft. In addition to the specific information on grade-level standards shown in the table, students, according to the draft standards, will be expected to develop practices that connect to scientific practices (the ability to reason abstractly and quantitatively, construct arguments and critique the reasoning of others, model with mathematics, and use appropriate tools strategically) (Common Core State Standards Initiative 2010; *www.corestandards.org/the-standards/mathematics*).

CHAPTER 3: The Framework and Instructional Tools at the Elementary Level

Figure 3.6

Ms. Hernandez Implements Learning Target #1 ("Food is a substance that provides fuel and building materials for organisms.")

Ms. Hernandez is a fifth-grade science teacher at Mesa Elementary School. She is teaching a lesson for the first time. The lesson is focused on the role of food for organisms. She has already worked with students on their definitions of *food* and wants to see how much they understand about what happens to food when it is eaten by an animal. She shows a concept cartoon (Figure 2.11, p. 69) related to the concept and asks students to write entries in their science notebooks that explain which cartoon they think provides the best answer to the question. They are to include thorough rationales for their answers. She suggests that they draw images as well, but their drawings must be annotated and must include at least a paragraph that explains their reasoning.

Once students complete their individual work, she places them in groups of four, asking that they share their responses. She asks them to reach consensus on the best answer, reminding them to use their listening skills and make sure each member of their group contributed to the dialogue. Once groups complete their discussions, Ms. Hernandez facilitates a whole-class discussion about their thinking. During the discussion, she periodically asks students to rephrase something a peer said and also asks several students to summarize the group's thinking to that point. She urges students to share questions they still have and she lists these questions on chart paper. She identifies a question from the list, "Does what an animal eats become part of its body?" and tells them that tomorrow they will design experiments to determine the answer to this question.

Before school the next day, Ms. Hernandez sets up multiple supply tables: one with a variety of organisms, a second with a variety of measuring devices, a third with various materials used to feed and shelter the organisms, and the last with a range of text materials (e.g., several textbooks other than the ones in use by students, science magazines, and brochures on the care of the various organisms). Today students are to (1) choose a type of organism to test the question, (2) select appropriate resource materials and access the internet to determine the best way to care for the organisms during their experiments, and (3) outline in their science notebooks thorough plans for conducting their experiments. She reminds them that it is important to choose appropriate measurement devices and units for answering the question. As students work, she goes from group to group checking their plans. Once the plans are approved, groups set up appropriate habitats, make initial measurements, and clean up. Ms. Hernandez reminds them to record in their notebooks a claim and their reasoning for it.

Over the course of the next two weeks, students implement their plans. Each day they make careful measurements (using the appropriate metric measurement), log all data using well-designed tables, and include thorough written descriptions of their observations and conjectures. Once all data are collected, the groups discuss and analyze the data, drawing implications. They once again use available text and online resources to help them write explanations based on the evidence they gathered. They are asked to include whether or not the evidence supported their claims. If so, they should explain how; if not, they should explain why and write their best possible explanations based on what they learned.

Finally, small groups share their results with the entire class. They prepare posters that include their questions, claims, procedures, data (including tables and graphs), and their explanations. Ms. Hernandez orders the group presentations in a sequence that she feels will slowly build the understandings of the class. Student groups listen carefully during the presentations and ask questions of their peers. Ms. Hernandez probes students' thinking about the effectiveness of their experimental designs, measurements, and explanations. Students are periodically asked to rephrase their comments after their presentations and then to summarize what they understand at that point. Ms. Hernandez closes with a summary of their findings.

CHAPTER 3: The Framework and Instructional Tools at the Elementary Level

Table 3.5
Mathematics Standards Supported by Science Instruction

Mathematics Standard	Expectations of the Mathematics Standard	Science Connections
Numbers and Operations: Understand numbers, ways of representing numbers, relationships among numbers, and number systems	Understand the place-value structure of the base-ten number system and be able to represent and compare whole numbers and decimals	Using the metric system hinges on students' understandings about base-ten and decimals.
Algebra: Understand patterns, relations, and functions	• Describe, extend, and make generalizations about geometric and numeric patterns • Represent and analyze patterns and functions using words, tables, and graphs	These expectations align with analyzing data; representing that data in graphs, tables, and equations; and drawing conclusions in inquiry situations.
Algebra: Use mathematical models to represent and understand quantitative relationships	Model problem situations with objects and use representations such as graphs, tables, and equations to draw conclusions	
Algebra: Analyze change in various contexts	• Investigate how change in one variable relates to change in a second variable • Identify and describe situations with constant or varying rates of change and compare them	
Geometry: Specify locations and describe spatial relationships using coordinate geometry and other representational systems	• Make and use coordinate systems to specify locations and to describe paths • Find the distance between points along horizontal and vertical lines of a coordinate system	Coordinate systems are important for mapping and satellite imagery. They are essential to understand graphing.

Table 3.5 *(continued)*

Mathematics Standard	Expectations of the Mathematics Standard	Science Connections
Geometry: Use visualization, spatial reasoning, and geometric modeling to solve problems	• Create and describe mental images of objects, patterns, and paths • Identify and build a three-dimensional object from a two-dimensional representation • Identify and draw a two-dimensional representation of a three-dimensional object • Recognize geometric ideas and relationships and apply them to other disciplines and to problems that arise in the classroom or in everyday life	These are all essential to modeling.
Measurement: Understand measurable attributes of objects and the units, systems, and processes of measurement	• Understand such attributes as length, area, weight, volume, and size of angle and select the appropriate type of unit for measuring each attribute • Understand the need for measuring with standard units and become familiar with standard units in the customary and metric systems • Carry out simple unit conversions, such as from centimeters to meters, within a system of measurement • Understand that measurements are approximations and how differences in units affect precision	These same understandings are essential to measurement in science.
Measurement: Apply appropriate techniques, tools, and formulas to determine measurements	Select and apply appropriate standard units and tools to measure length, area, volume, weight, time, temperature, and the size of angles	Selecting appropriate units and tools is equally important in science.
Data Analysis/ Probability: Formulate questions that can be addressed with data and collect, organize, and display relevant data to answer them	• Design investigations to address a question and consider how data-collection methods affect the nature of the data set • Collect data using observations, surveys, and experiments • Represent data using tables and graphs such as line plots, bar graphs, and line graphs • Recognize the differences in representing categorical and numerical data	These are essential skills during experimental inquiry.

(continued)

CHAPTER 3: The Framework and Instructional Tools at the Elementary Level

Table 3.5 *(continued)*

Mathematics Standard	Expectations of the Mathematics Standard	Science Connections
Data Analysis/ Probability: Select and use appropriate statistical methods to analyze data	• Describe the shape and important features of a set of data and compare related data sets, with an emphasis on how the data are distributed • Use measures of center, focusing on the median, and understand what each does and does not indicate about the data set • Compare different representations of the same data and evaluate how well each representation shows important aspects of the data	These standards are all essential to scientific inquiry, specifically when we instruct students to give priority to evidence in responding to questions, formulate explanations from evidence, and communicate and justify explanations.
Data Analysis/ Probability: Develop and evaluate inferences and predictions that are based on data	Propose and justify conclusions and predictions that are based on data and design studies to further investigate the conclusions or predictions	
Data Analysis/ Probability: Understand and apply basic concepts of probability	• Describe events as likely or unlikely and discuss the degree of likelihood using such words as certain, equally likely, and impossible • Predict the probably of outcomes of simple experiments and test the predictions • Understand that the measure of the likelihood of an event can be represented by a number from 0 to 1	
Problem Solving	Solve problems that arise in mathematics and in other contexts	
Connections	Recognize and apply mathematics in contexts outside of mathematics	
Representation	Use representations to model and interpret physical, social, and mathematical phenomena	

CHAPTER 3: The Framework and Instructional Tools at the Elementary Level

Table 3.6
Common Core State Standards for Mathematics (Grades 3–5) With Connections to Science

	Summary of Common Core State Standard	**Sample Science Connections**
3rd Grade	Represent and Interpret Data—Draw scaled picture graphs and bar graphs to represent data sets; generate measurement data and show the data making a line plot	Representing and interpreting data tie in nicely to any experimental work your students might do.
	Geometric Measurement—Measure areas in square units	This standard could be tied to work in the field when students measure areas of study and collect data in those areas.
	Problem solving (included across the standards) related to measurement and estimation of time intervals, liquid volumes, and masses of objects	These standards also relate in multiple ways to experiments your students might do.
4th Grade	Represent and Interpret Data—Make a line plot to display a set of measurements in fractions of a unit	
	Recognize and measure angles; draw points, lines, segments, range, angles	Many geometric skills are applicable to physics-related topics such as force and motion.
	Problem solving (included across the standards) related to measurement and estimation of time intervals, liquid volumes, and masses of objects	These standards also relate in multiple ways to experiments your students might do.
5th Grade	Understand place values, including writing decimals to the thousandths place	These standards are valuable to help students understand metric measurement and apply it during experiments and fieldwork.
	Convert like measurement units within a measurement system, using conversions to solve real-world problems	
	Represent and Interpret Data—Make a line plot to display a data set of measurements in fractions of a unit	These standards also relate in multiple ways to experiments your students might do.
	Recognize volume as an attribute of solid figures; understand concepts of volume measurement; solve real-world problems involving volume	
	Solve real-world problems graphing points on a coordinate plane	

Hard-to-Teach Science Concepts

CHAPTER 3: The Framework and Instructional Tools at the Elementary Level

Your students all have mental models they use to explain observed phenomena. Some of their models reflect naive conceptions of the world, and you need to help them reconstruct their models to better reflect scientific thinking. They can use mathematics to model objects, relationships, and events (see Instructional Tool 2.7, pg. 107). You also want them to understand that numbers and shapes help describe the world and predict things about it (AAAS 2001b). For example, if you want students to learn about how food is used in an organism, you want them to see that the food impacts the organism's size. Unless they gather data through measurement and then graph those data, it is difficult for them to conceptualize the information and use it to predict what might happen to the organism if food were limited. The integration of mathematics and science helps you instruct students on how to develop models and interpret models so they can better understand their world.

Science as Related to Literacy

Reform efforts for language literacy consider reform to be a three-fold process—meaning making, comprehending the meaning of other people, and communicating meaning to others. Language is seen as a powerful tool to present our ideas to the world and receive ideas from the world (Hammerman and Musial 2008). The first two chapters of this book provide ways to uncover students' ideas and to collaborate with others to make sense of science phenomena. Language literacy processes fit in beautifully with those goals. Reading, writing, speaking, and listening are essential to scientific inquiry and can be used during engagement and exploration activities, designing and conducting investigations, analyzing and interpreting data, and presenting findings and understandings (Century et al. 2002). Inquiry-based science classrooms can provide a real-world context for developing language literacy.

Language literacy is central to conceptual change and is reflected in the Instructional Planning Framework itself and in the Instructional Tools focused on linguistic representations (Tool 2.4 on Writing to Learn, Tool 2.5 on Reading to Learn, and Tool 2.6 on Speaking to Learn). We do not suggest that you purposefully teach skills in reading, writing, and speaking in your science lessons but rather that you expect students to use those skills in learning science concepts. Likewise, the strategies in the Instructional Tools focus on science comprehension, although they depend on reading, writing, and speaking. Overall, science as a content focus provides a real-world context in which to enhance your students' understandings and abilities in both science and literacy.

Instructional Tools 2.4–2.6 address the use of language in learning science, so that topic is not extensively covered here. However, see Table 3.7 for some key connections between the two disciplines. (Rather than use the various language arts standards, the authors used the draft Common Core Standards for the English Language Arts. Go to the Common Core homepage at *www.corestandards.org* and select the "English Language Arts Standards" link to find the source for the information in Table 3.7).

Table 3.7

Science Connections to the English Language Arts

Language Arts Area	Specific Standards Area	Brief Description of Relevant Fifth-Grade Expectations	Science Connections
Reading Standards for Informational Text	Key Ideas and Details	Support statements about text with quotes; determine how text supports main ideas; summarize text; explain the relationship between scientific concepts using information from one or more texts.	Each of these expectations connects with some of the features of scientific inquiry: give priority to evidence in responding to questions, formulate explanations from evidence, and connect explanations to scientific knowledge.
	Craft and Structure	Determine meanings of domain-specific words relevant to a fifth-grade subject area; analyze two accounts of the same topic and summarize similarities and differences between the two.	
	Integration of Knowledge and Ideas	Use multiple sources to locate an answer to solve a problem; identify an author's use of evidence to support claims in text and identify the evidence that supports the claim; integrate information from various texts to write or speak about a subject.	
Writing Standards	Text Types and Purposes	Write opinions and write informative/explanatory pieces.	Students write opinions in science when dealing with issues related to science content they study. Explanatory pieces are essential in writing scientific explanations.
	Research to Build and to Present Knowledge	Complete brief, focused research to build knowledge using various sources; gather relevant information from experience and print/digital sources and summarize information in written work; write in response to informational sources, drawing evidence to support analysis.	Research is essential in science and gathering information based on experience (inquiry) and resources helps students to build explanations and support those explanations with evidence.

(continued)

CHAPTER 3: The Framework and Instructional Tools at the Elementary Level

Table 3.7 *(continued)*

Language Arts Area	Specific Standards Area	Brief Description of Relevant Fifth-Grade Expectations	Science Connections
Speaking and Listening Standards	Comprehension and Collaboration	Initiate and engage in discussions, coming prepared to draw from sources about the topic of discussion, contribute comments that build on the ideas of others, ask questions to clarify or follow up on ideas, and draw conclusions based on the discussion; summarize ideas and details presented orally, visually, graphically; summarize claims made by a speaker and evidence used to support the claims.	Discourse is essential when students collaborate to make sense of science experiences. It is also important that they can effectively summarize and present their ideas.
	Presentation of Knowledge and Ideas	Present an organized and logical report using appropriate and specific facts and details to develop main ideas, using visual displays as appropriate.	Communicating results is core to doing science.

    The language arts abilities described in Table 3.7 are essential in the instructional unit on the flow of matter and energy in ecosystems (refer to Learning Target #1 in Table 2.15, p. 60). Students use their speaking and listening skills when working with partners, in their small groups, and during whole-class discussions. They must initiate and engage in discussion, build on and clarify ideas, and build their explanations based on sources and experiences. Students also use their speaking and listening skills during group presentations to communicate results, as outlined in Table 3.6. In addition, these skills are critical to learning during the Gallery Walk of concept maps as the teacher facilitates discussion.

    Reading and writing are also essential to multiple aspects of the instructional plan. Science notebooks are used to respond to the concept cartoon, when students select the responses they see as most accurate and write rationales for their choices. Eventually, they establish a group claim that is recorded in their notebooks and they substantiate their thinking through concept mapping. These concept maps are used and revised throughout their work, providing a framework for their thinking and explanations, written and oral.

    Students use various digital and print resources to compare their explanations to more scientific ones. They use their notes written during experimentation, along with quantitative data, to summarize written conclusions and share the conclusions with the class. They use print and digital resources to build explanations based on evidence they gathered and return at the end of the lesson to the concept cartoon, writing individual responses and their explanations for the choice.

Nonfiction science texts and digital resources and writing are used as they are used by scientists—to learn what is known, to understand perspectives of others, and to use these perspectives to develop their own understandings. Fiction can be used to engage students in initial thinking about a concept; nonfiction texts are usually used at the end of an investigation to confirm or extend what the students have learned. Science fiction can be used for students to learn about separating scientific fact from fiction. You can also have students use their newly gained science understandings to write creatively about science. Regardless of selected resources, reading and writing are essential to science learning—and science provides the ideal context for meaningful reading and writing.

Fiction and nonfiction resources related to the flow of matter and energy in ecosystems are found in the Build Your Library section on page 154. Consider this section a starting place and continue to hunt for effective resources. Each of the chapters in Part II of this book also includes a Build Your Library section.

Variations in Third, Fourth, and Fifth Grades

The work in chapters 2 and 3 described an instructional unit on the flow of matter and energy in ecosystems targeted specifically for use in fifth grade. Although this is a common grade level at which this topic is taught, you may be teaching it at a different level. The plan as written would likely serve well at the fourth-grade level. But if your school addresses this content in third grade, you might have to modify it slightly, in one or more of the following ways:

1. Consider modification of the metacognitive focus and/or provide increased scaffolding because students' abilities to plan and self-regulate may not be as developed. For instance, you might scaffold students' completion of the organizer for fieldwork (Table 2.16 on p. 71) by outlining the tasks in column #1 during a whole-class discussion. Or you might shift the metacognitive focus to something a little less challenging, such as "identify what you know and don't know" (see Instructional Tool 2.2 on p. 84). Consider more-structured inquiry, especially if the unit is taught early in the school year. Further, help students in their efforts to observe and gather data, organize those data in meaningful ways, and summarize findings.
2. Modify reading materials and online resources used for student content resources.
3. Definitely limit content coverage to food chains. At the fifth-grade level, students are generally able to understand the complexity of webs and the interactions of systems, although deep understandings are difficult even at that level. For third graders, you should definitely limit coverage to simple food chains.

CHAPTER 3: The Framework and Instructional Tools at the Elementary Level

Topic: Food Webs (K–4)
Go to: www.scilinks.org
Code: HTT001

Topic: Food Chains (K–4)
Go to: www.scilinks.org
Code: HTT002

Topic: Matter and Energy (K–4)
Go to: www.scilinks.org
Code: HTT003

Topic: Matter and Energy (5–8)
Go to: www.scilinks.org
Code: HTT004A

Topic: Foods and Energy (5–8)
Go to: www.scilinks.org
Code: HTT004B

Topic: How Do Organisms Get Energy? (5–8)
Go to: www.scilinks.org
Code: HTT018

Build Your Library

For Your Students

NSTA Outstanding Science Trade Books for Students

Note: These books relate only to the flow of matter and energy in ecosystems.

Book about Food Chains (Magic School Bus)
The Magic School Bus Gets Eaten: A Food Chain Frenzy (Magic School Bus Chapter Book #17)
Guts: Our Digestive System (Seymour Simon)
Here Is the Tropical Rain Forest (Madeleine Dunphy)
Horseshoe Crabs and Shorebirds: The Story of a Food Web (Victoria Crenson)
Predators (John Seidensticker and Susan Lumpkin)
The Case of Mummified Pigs: And Other Mysteries in Nature (Susan E. Quinlan)
The Temperate Forest: A Web of Life (Philip Johansson)
The Wolves are Back (Jean Craighead George)
When the Wolves Returned: Restoring Nature's Balance in Yellowstone (Dorothy Hinshaw Patent)
Who Eats What? Food Chains and Food Webs (Patricia Lauber)

Magazines and Online Resources

National Geographic Kids (Print and online at *http://kids.nationalgeographic.com/kids*)
Odyssey (Print and online at *www.odysseymagazine.com*)
Ranger Rick (Print and online at *www.nwf.org/Kids/Ranger-Rick.aspx*)
Science News for Kids (*www.sciencenewsforkids.org*)

For You

Text Resources: Differentiation

- *Differentiated Instructional Strategies for Science, Grades K–8* (Gregory and Hammerman 2008). This book discusses specific strategies for (1) activating and engaging, (2) acquiring and exploring, (3) explaining, applying, and creating meaning, and (4) elaborating and extending. These categories tie in nicely with any learning cycle, including the Instructional Planning Framework and the 5E Instructional Model. Examples of learning and science-interest inventories are also included.
- *Integrating Differentiated Instruction and Understanding by Design* (Tomlinson and McTighe 2006). If your school and/or district focuses on UbD (Understanding by Design), you will find this excellent book particularly helpful. It combines UbD with differentiation (à la Carol Tomlinson), providing clear descriptions as well as lesson ideas in the various disciplines.
- *The Parallel Curriculum: A Design to Develop High Potential and Challenge High-Ability Learners* (Tomlinson et al. 2002). This book provides rich resources and ideas for differentiation. Consider focusing on Chapters 2 and 3.

- *Differentiating for the Young Child: Teaching Strategies Across the Content Areas, PreK–3* (Smutny and Von Fremd 2010). This book provides practical examples and guidelines for differentiation, not only in science but for other disciplines as well.
- *Teaching Constructivist Science: Nurturing Natural Investigators in the Standards-Based Classroom* (Bentley, Ebert II, and Ebert 2007). This book includes practical applications, teaching strategies, activities, and assessment tools.

Text Resources: Interdisciplinary Approaches to Teaching Science

- *Activities Linking Science with Math, K–4* (Eichinger 2009a) and *Activities Linking Science with Math, 5–8* (Eichinger 2009b) are hands-on guides for preservice and inservice elementary and middle school teachers who want to connect science instruction with other areas of study, including visual arts, social sciences, language arts, and especially math.
- The *Everyday Science Mysteries: Stories for Inquiry-Based Science Teaching* series (Konicek-Moran 2008, 2009, 2010, 2011) provide wonderful mystery stories with student misconceptions embedded in the text. The books in the series require reading and writing to write the final chapter and expect students to act as sleuths to find answers.
- *Integrating Science with Mathematics and Literacy: New Visions for Learning and Assessment* (Hammerman and Musial 2008) builds a rationale for and provides examples of performance assessments that integrate science, mathematics, and literacy.
- *Outdoor Science: A Practical Guide* (Rich 2010) shows teachers how to set up an outdoor classroom. It provides lessons in various areas of science and discusses how to integrate mathematics, social studies, and language arts.
- *Questions, Claims, and Evidence: The Important Place of Argument in Children's Science Writing* (Hand et al. 2008) describes steps to link literacy and inquiry.
- *Science and Literacy: A Natural Fit* (Worth et al. 2009) helps teachers make connections between balanced literacy instruction and experiential science, providing examples that show how to make talk and writing essential parts of science inquiry.
- *Seeds of Science/Roots of Reading* (Lawrence Hall of Science and the University of California at Berkeley 2010) is a fully integrated science and literacy program intended to give students opportunities to investigate science topics through inquiry and to use reading, writing, and discourse to make sense of their inquiry experiences and reflect on them. Though the book was developed as a full curriculum, individual books and associated strategy guides are available for purchase.
- *Teaching Reading in Science: A Supplement to Teaching Reading in the Content Areas Teacher's Manual.* 2nd ed. (Barton and Jordan 2001) includes a great selection of strategies to support vocabulary development, negotiating informational text, and reflection.

- The *Uncovering Student Ideas in Science* series (Keeley 2011; Keeley, Eberle, and Farrin 2005; Keeley, Eberle, and Tugel 2007; Keeley, Eberle, and Dorsey 2008; Keeley and Tugel 2009; Keeley and Harrington 2010; for information on the *Uncovering Student Ideas in Science* series, go to *www.nsta.org/publications/press/uncovering.aspx*) includes various probes that require initial written responses from students but also open doors to partnered and small-group discourse.
- *Science Notebooks: Writing About Inquiry* (Campbell and Fulton 2003) discusses how to use notebooks in support of student thinking and learning.
- *Writing in Science: How to Scaffold Instruction to Support Learning* (Fulwiler 2007) provides specific guidelines, strategies, activities, and student work examples you can use to implement a science notebook system.

Online Resources
- 50 Strategies for Teaching English Language Learners: *http://edweb.sdsu.edu/people/jmora/pages/50strategies.htm*
- Tools for Differentiation: *http://toolsfordifferentiation.pbworks.com*
- Differentiation Strategies: *www.saskschools.ca/~bestpractice*
- Tiered Curriculum Project: *www.doe.in.gov/exceptional/gt/tiered_curriculum/welcome.html*
- Science Notebooks in K–12 Classrooms: *www.sciencenotebooks.org*
- Connecting Elementary Science and Literacy: *http://cse.edc.org/products/sciencelieteracy/matrixhome.asp*
- The SCENE website: *http://scene.asu.edu/habitat/inquiry.html#4*. See the overview of a questioning cycle in inquiry and a questioning tree (Figure 3 on that site).
- Seeds of Science/Roots of Reading website: *http://seedsofscience.org*

Endnotes

[1] Recall the five natural learning systems (Given 2002; Gregory and Hammerman 2008) that were mentioned in the introduction to this book: the cognitive system, the emotional system, the social system, the reflective (or metacognitive) system, and the physical system. The first two chapters of the book focused on cognition and metacognition. This chapter focuses on the emotional and social learning systems required to establish a safe learning environment and to engage each student in the learning process. The final system (physical) is addressed indirectly through the active learning typical of inquiry and directly through Instructional Tool 2.10: Kinesthetic Strategies, p. 125.

[2] These ideas are drawn from *Integrating differentiated instruction and understanding by design* (Tomlinson and McTighe 2006).

PART II
Toolbox Implementation: Using the Framework and Instructional Tools With Hard-To-Teach Science Topics

Chapter 4

Matter and Its Transformations: Gas Is Matter

CHAPTER 4: Matter and Its Transformations: Gas Is Matter

Overview

A cube of ice, water in a cup, and air in an inflated balloon—what makes each of these forms of matter different? How do we help students understand that these forms of matter can be transformed so that they are better able to move from their concrete observations to more deeply understand what they cannot see? In other words, how do we help students move from macro-level concepts to the micro-level concepts of matter and its transformations? How do we enable students to really understand that matter is anything that has weight and occupies space?

One summer while working with a group of 5- to 10-year-old children at a day care center, I asked them, "What is in this inflated balloon?" A young girl immediately said, "Air." I could not help but wonder what she and the others really understood about this stuff called air. Did they associate air with the term *gas*, and did they consider gas as another form of matter? On another day, I asked the same children, "What is all around us?" A third grader said that air was all around us. I continued with, "What do we breathe?" Immediately another child said, "Oxygen." These children were all aware of something invisible that can be named, but did they understand that this something is a form of matter—a gas?

I recently observed a woman filling a cylinder at the local gas station. She parked her car, went to the air pump, and connected it to her red cylinder. She squatted down for at least 10 minutes as the connection allowed "stuff" to be sent through the pump to her red cylinder. I approached the woman and inquired what she would use the "stuff" from the air pump for. I told her that I teach science and was working on a states-of-matter project. She smiled and told me that she used the "stuff" in the red cylinder to fill her low tires. I wondered how many students had seen their parents use these air pumps and made the connection to gas and matter, a practical application of the content they were learning.

Why This Topic?

The foremost reason this topic was chosen for inclusion in this book is that a lot of elementary science instructional time in grades 3–5 is spent studying solids, liquids, and gases. A second reason is that the concept of matter and its transformations (matter going from one state to another) is abstract for third through fifth graders, according to my many exchanges with elementary science teachers and elementary science coaches in the Omaha Public Schools, as well as research on elementary science teaching (Sere 1985). Many elementary-level students can recite the standard definition of *matter:* Matter is anything that has weight and takes up space. However, few students are able to explain matter in a way that provides a conceptual foundation for learning in later grades when they are introduced to molecules and position of molecules in each of the three states.

I'm not sure whether the five- to ten-year-old children I mentioned earlier would say that air has mass, but they understood that it surrounds us. Matter is a difficult

CHAPTER 4: Matter and Its Transformations: Gas Is Matter

topic because students are not able to visualize the micro parts that make up different types of matter; therefore, they form a concrete model from macro examples. Students in grades 3–5 lack an appreciation and understanding of the tiny particles that make up matter and thus attribute the macroscopic properties to the particles (AAAS 2001b). Further, students believe that there must be something in the space between particles. They have difficulty appreciating the intrinsic motion of particles in solids, liquids, and gases and have problems conceptualizing forces between particles (CLIS 1987). Finally, because of the complexity of the pieces and levels of organization of matter, matter and its transformations are conceptually difficult for students in grades 3–5.

However, it is possible to help students more deeply understand matter and its transformations—in particular, the third phase: gas. With special instruction, some students in fifth grade can identify that air is the final location of evaporating water (Russell and Watt 1990), but they must first accept air as a permanent substance (Bar 1989), which is a challenging concept for upper elementary students (Sere 1985).

The Predictive Phase

It is important that our students have a conceptual understanding about matter and its forms rather than just the ability to recite phrases such as "matter is anything that has mass and takes up space." Using the Instructional Planning Framework described in Part I of this book promotes instruction that can move students' learning from rote memorization to understanding concepts through the inquiry process.

Stages I–V

(To see Stages I–V, go to Table 2.1, p. 19, Ten-Stage Process for Implementing the Instructional Planning Framework.)

For my lessons in this chapter (pp. 166–169), I used *Benchmarks for Science Literacy* (AAAS 1993) and *National Science Education Standards* (NRC 1996) to identify the standards related to matter and its transformations. Table 4.1, page 162, identifies the targeted standards and those in earlier grades that are foundational to understanding this topic. I then "unpacked" the standards into smaller chunks, which included the various embedded concepts, to identify smaller learning targets designed to lessen the gap between the students' understanding and the scientific explanations presented. This process led to the three learning targets and essential understandings shown in Figure 4.1, page 163.

Students must understand that the "stuff" we call *gas* is one form of matter and that this "stuff" is formed when water, a liquid, evaporates and goes into the air. Finally (and possibly the most misunderstood idea), students must understand that this gas is made of a permanent substance and it does not disappear. The essential understanding is that a gas is a form of matter, just like liquids and solids. A gas can change state and become a liquid or solid, but the matter never disappears.

CHAPTER 4: Matter and Its Transformations: Gas Is Matter

Table 4.1

National Science Education Standards and Benchmarks for Science Literacy Related to "Matter and Its Transformations"

	National Science Education Standards
Grades K–4	Objects have many observable properties, including size, weight, shape, color, temperature, and the ability to react with other substances. Those properties can be measured using tools, such as rulers, balances, and thermometers. (p. 127)
	Objects are made of one or more materials, such as paper, wood, and metal. Objects can be described by the properties of the materials from which they are made, and those properties can be used to separate or sort a group of objects or materials. (p. 127)
	Materials can exist in different states—solid, liquid, and gas. Some common materials, such as water, can be changed from one state to another by heating or cooling. (p. 127)
	Benchmarks for Science Literacy
Grades 3–5	Heating and cooling cause changes in the properties of materials. Many kinds of changes occur faster under hotter conditions. (p. 77)
	No matter how parts of an object are assembled, the weight of the whole object made is always the same as the sum of the parts; and when a thing is broken into parts, the parts have the same total weight as the original thing. (p. 77)
	Materials may be composed of parts that are too small to be seen without magnification. (p. 77)

Using the learning targets, I retained essential ideas and eliminated unnecessary vocabulary from the Standards and Benchmarks. I added the essential understandings that are prerequisite instruction to the learning targets and necessary vocabulary to the planning template for the predictive phase (Table 4.2). (The identified skills and the criteria to demonstrate understanding included in Table 4.2 are explained later in this chapter.)

CHAPTER 4: Matter and Its Transformations: Gas Is Matter

Figure 4.1

Learning Targets and Sequence for "Gas Is Matter"

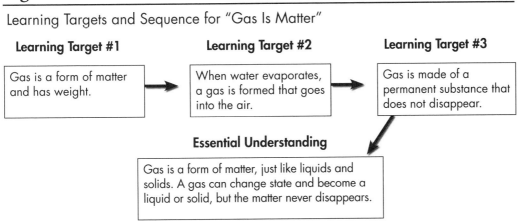

Learning Target #1: Gas is a form of matter and has weight.

Learning Target #2: When water evaporates, a gas is formed that goes into the air.

Learning Target #3: Gas is made of a permanent substance that does not disappear.

Essential Understanding: Gas is a form of matter, just like liquids and solids. A gas can change state and become a liquid or solid, but the matter never disappears.

Table 4.2

Predictive Phase: Planning Template for "Gas Is Matter"

Lesson Topic: "Gas Is Matter"	
Essential Understandings	Gas is a form of matter, just like liquids and solids. A gas can change state and become a liquid or solid, but the matter never disappears.
Knowledge Required From Previous Instruction	• Objects have many observable properties including size, weight, shape, color, temperature, and the ability to react with other substances. Those properties can be measured using tools, such as rulers, balances, and thermometers. • Objects are made of one or more materials, such as paper, wood, and metal. Objects can be described by the properties of the materials from which they are made, and those properties can be used to separate or sort a group of objects or materials. • Materials can exist in different states—solid, liquid, and gas. Some common materials, such as water, can be changed from one state to another by heating or cooling. • Heating and cooling cause changes in the properties of materials. Many kinds of changes occur faster under hotter conditions. • No matter how parts of an object are assembled, the weight of the whole object made is always the same as the sum of the parts; and when a thing is broken into parts, the parts have the same total weight as the original thing.
Knowledge and Skills to Be Learned	**Concepts:** See the Learning Targets. **Vocabulary:** matter, solid, liquid, gas, space, weight, mass, vapor **Skills:** • Measure mass using different kinds of scales • Manipulate science equipment

(continued)

Hard-to-Teach Science Concepts   163

CHAPTER 4: Matter and Its Transformations: Gas Is Matter

Table 4.2 (continued)

Criteria to Demonstrate Understanding	• Compare the masses of an uninflated and inflated balloon and explain what this has to do with matter • Thoroughly describe the difference between a liquid and gas • Draw what happens when a liquid such as water is boiled • Check data for accuracy during experimentation (inquiry focus) • Establish a personal learning goal for the unit of instruction (metacognitive focus)
Learning Targets and Instructional Plans	
Target #1	Gas is a form of matter and has weight.
	Instructional Plan for Learning Target #1: Will be completed during the responsive phase.
Target #2	When water evaporates, a gas is formed that goes into the air.
	Instructional Plan for Learning Target #2: Will be completed during the responsive phase.
Target #3	Gas is made of a permanent substance that does not disappear.
	Instructional Plan for Learning Target #3: Will be completed during the responsive phase.

I used Instructional Tool 2.1 (p. 76) to identify the essential feature of inquiry on which these lessons would focus. Because of the abstract nature of the content, I chose a more structured inquiry approach, one in which I provide the questions and some of the procedures. I want students to formulate explanations from the evidence that they gather. These explanations will show me if they understand the abstract concepts. I also established a criterion to determine understanding of this inquiry goal—*formulate explanations from evidence*—and added it to Table 4.3, p. 167.

I again used Instructional Tool 2.2 (p. 83), this time to identify the metacognitive focus for these lessons. I selected a self-regulated thinking strategy: *identify what you know and don't know*. I then added the criterion to determine understanding of this metacognitive goal to Table 4.3, p. 167.

The Responsive Phase

I worked through the process outlined in Chapter 2 to select research-based strategies to use along with common activities that help represent phenomena associated with my three learning targets. Students must be able to explain what they think about phenomena and change their thinking based on what they experience. They should explain their ideas to one another in pairs and in groups and then to the teacher. Students should not be rushed through curricula by, for example, simply memorizing definitions; instead, they should have time to study the phenomena through investigations and discussions. The learning targets and strategies that a teacher selects should allow students to think about what they observe, ask questions about *why* they see what they see, and then investigate and talk among themselves about their observations during each of the lessons. The next section describes strategies to use with relatively common activities that cover the learning targets.

Stage VI: Research Children's Misconceptions Common to This Topic That Are Documented in the Research Literature

Part I of this book outlined a process to identify research-identified misconceptions that students have. I followed this process (Figure 2.6, p. 36) and identified the misconceptions listed in Figure 4.2.

Figure 4.2

Research-Identified Misconceptions About "Gas Is Matter"

Some common misconceptions about gas as matter:
1. When water evaporates, it ceases to exist.
2. Water changes its location but remains a liquid OR water is transformed into some other perceptible form (e.g., fog, steam, droplets).
3. Air is not a permanent substance.
4. Air is not matter because we cannot see it.
5. Air does not have mass.
6. Gas means gasoline.
7. Gases just disappear.

Over the past 20-plus years, researchers have documented students' misconceptions about matter, in particular the gas phase. Students in grades 3–5 have difficulty accepting that evaporated water is still present in the air and that a gas is matter, primarily because they cannot see either form. Further, if they cannot see gas, how can it have mass? Students must be given opportunities to challenge these misconceptions so that they can understand that the "stuff" we call gas is really matter.

Stage VII: Select Strategies to Identify Your Students' Preconceptions

Several weeks before teaching the lessons on matter, I read Instructional Tool 2.3 (p. 42) and determined that I would use an informational text strategy to identify student preconceptions and use the information I gathered to help design my lesson. Rather than using the familiar KWL strategy (what do I **K**now, what do I **W**ant to know, and what have I **L**earned?), I used a similar strategy called THC (what do you **T**hink, **H**ow can we find out, and what do we Conclude?). This strategy aligns nicely with the way scientists work (Crowther and Cannon 2004). At this point I use only the "think" portion of THC. I might ask a question such as, "What do you think happens when an ice cube melts or when a puddle disappears?" Student responses help me understand how to design my lesson. This strategy was added to the Strategy Selection Template (Table 4.3).

Stage VIII: Select Strategies to Elicit and Confront Your Students' Preconceptions

I continued to use the process outlined in Part I to select strategies to elicit and confront student preconceptions. For example, to elicit students' preconceptions for Learning Target #2, I planned to use drawings of what students believe is happening when solid ice cubes melt and then evaporate. In the lesson, students will observe water as it evaporates and draw what they believe happened, explaining why no more water is visible. When students share their drawings, their conceptions might be confronted by the different ideas of their peers. I also selected hands-on experiments as a strategy to further confront students' preconceptions. Once I had selected strategies for each learning target, I added them to the Strategy Selection Template (Table 4.3).

Stage IX: Select Sense-Making Strategies

I continued to use the process from Part I to select sense-making strategies. Once I had made my selections for each learning target, I added them to the Strategy Selection Template (Table 4.3). In Lesson 2, students will be expected to refine their explanations and compare them to scientific explanations. Working with their explanations in this way allows the students to further clarify their thinking about the concepts. (It is informative for the teacher, too.)

Stage X: Determine Formative Assessments

I again used the process outlined in Part I, this time to select assessment strategies. For Lesson 2, students' annotated drawings and explanations during class discussion will serve as formative assessments. Once I selected strategies for each learning target, I added them to the Strategy Selection Template (Table 4.3).

The Lessons: Teaching and Learning About "Gas Is Matter"

Lesson 1

In this lesson, students design and conduct an investigation in which they place a large, deflated balloon[*] on a digital balance and record its mass. A digital balance is generally not part of the elementary classroom; the teacher can, however, borrow this type of balance from a middle or high school teacher. The exact mass that the digital balance provides makes it a necessary piece of equipment.

---

*Safety Note: Avoid using latex balloons. Some students may have allergic reactions to natural latex rubber.

Table 4.3

Strategy Selection Template for "Gas Is Matter"

	Learning Target #1	**Learning Target #2**	**Learning Target #3**
	Gas is a form of matter and has weight.	When water evaporates, a gas is formed that goes into the air.	Gas is made of a permanent substance that does not disappear.
Possible Strategies for:			
Identifying Preconceptions	Informational Text Strategies: Use the THC strategy (what do you think, how can we find out, and what do we conclude?).		
Eliciting and Confronting Preconceptions	• Small- and large-group discourse • Hands-on experiments	• Annotated drawings • Small-group discourse • Hands-on experiments	• Large-group discourse • Manipulatives • Hands-on experiments
Sense Making	Students refine explanations and compare to scientific explanations	Students refine explanations and compare to scientific explanations	• Large-group discourse • Teacher questioning
Demonstrating Understanding	• Informal teacher questioning • Explanations based on evidence	• Large- and small-group discourse • Annotated drawings	• Written explanations
Selected Inquiry Strategy	Formulate explanations from evidence.		
Selected Metacognitive Strategy	Identify what you know and don't know.		

Next, students assign group members to carefully blow into the balloon, tie a knot at the end of the inflated balloon, and record the mass using the digital balance. Observe them closely as they blow into the balloon as some elementary children have difficulty with this task. (Also, it is unlikely, but possible, that a child could swallow the balloon, so remind the children to be careful.) If an air pump is available, use it. Have small and large groups of students describe any observed changes between the mass of the deflated balloon and that of the inflated balloon. If students note a change, ask them to describe the change. Ask them questions such as the following:

CHAPTER 4: Matter and Its Transformations: Gas Is Matter

- What is the change due to?
- What is in the balloon now that was not there before?
- What property of the "stuff" in the balloon can be identified?
- Do you think that scientists deal with this invisible "stuff" on a daily basis, and if so, how do they study the unseen "stuff"?

Lesson 2

The second lesson deals with another level of understanding about the topic "Gas Is Matter." This lesson transitions student thinking from the idea of gas as having matter and mass (Lesson 1) to ideas about the transformation of matter. Students observe a pile of ice that melts down to a liquid; over time, the liquid will evaporate into an invisible substance called a gas. Students draw what they believe happens to the liquid over time. This allows the students to explain in drawings their conceptual understanding of the phase change without naming it *phase change*.

This investigation takes more than a class period, but you must be patient and allow students to observe the phenomenon. In addition to having students draw what they predict will happen, ask students to share with one another their ideas about what is going on as the liquid water evaporates. Encourage debate among the students. Some possible questions that students can respond to are the following:

- What do you notice about the water that came from the ice cubes?
- Does it appear that the amount of the melted cubes is the same amount as the frozen cubes?
- How could we find out the answer to the previous question?
- What words can you use to describe what is happening to the water over time?
- Why do you think this is happening?

Students should have time to refine their explanations and compare them to scientific explanations. Have small- and large-group discussions; then ask students to make their final annotated drawings about what they understand.

Lesson 3

The final lesson is designed to further help students understand ideas developed during the previous lessons. Two soccer balls that are the same color and size are placed next to each other on the floor. Students make observations about the two balls as they pass them around and manipulate them. Ask students if the two balls feel the same. Does this mean that the balls have the same mass? Have two students weigh the balls and record the mass. Give students time to discuss their findings.

Ask the students what they think will happen if 10 pumps of air are inserted into one of the balls. Give students time to share their ideas and record them on a large wall Post-it. Have a student record the mass of the ball that received 10 pumps. Continue asking students questions about what they have observed. Example questions include

CHAPTER 4: Matter and Its Transformations: Gas Is Matter

- When you manipulate (hold, touch, squeeze) the balls, can you feel or hear anything from within the balls? Do you think there is something inside?
- If you think something is inside, how would you describe it?
- What is taking up the space in the balls?

Listen for explanations that lead to the idea that the mass change in the pumped ball is due to added gas from the pump. After a large-group discussion, ask students to write explanations of what they have observed.

Time for Reflection

Elementary teachers spend a lot of time teaching the concept of matter—solids, liquids, and gases. The study of this topic begins early in third grade and continues through sixth grade. The foundations about matter that are learned have a significant impact on students' understanding in later grades. Although students, according to research (Sere 1985), understand solids and liquids, they lack understanding about gases. The concept is abstract to elementary students, so it is very important to teach for conceptual understanding rather than memorization of vocabulary. Many types of investigations to teach this concept appear regularly in NSTA's member journal for elementary school teachers, *Science and Children* (e.g., using popcorn kernels and bubbles), but I chose very basic investigations that can be used with third- through fifth-grade students. I believe that these investigations will work well in helping students understand that gas is matter and that the "stuff" doesn't just disappear. Additionally, the questioning and discussions among students, along with the annotated drawings, will enable students to further understand the following concepts:

- Gases have mass.
- When the liquid phase becomes the gas phase, the amount of matter is still the same.

When students are allowed to observe, discuss, ask questions, manipulate objects, and explain in writing their ideas, they are building the foundation for further study about the molecular structure of matter in higher grades.

Take some time now to consider how these lessons might work in your school with your students. Reflect on the following questions:

- What aspects of these lessons would be easy to teach?
- Which lessons would be a challenge for you?
- Is there something suggested here that you have not tried before?

Topic: States of Matter (K–4)
Go to: *www.scilinks.org*
Code: HTT005

Topic: Properties of Matter (K–4)
Go to: *www.scilinks.org*
Code: HTT006

Topic: States of Matter (5–8)
Go to: *www.scilinks.org*
Code: HTT007

Topic: The Physical and Chemical Properties of Matter (5–8)
Go to: *www.scilinks.org*
Code: HTT008

CHAPTER 4: Matter and Its Transformations: Gas Is Matter

Ties to Literacy and Numeracy (Mathematics)

The use of trade books is a wonderful way to engage elementary students in reading about the states of matter, especially gases. A suggested list can be found in the Build Your Library section below. The use of vocabulary is another tie to literacy. A list of essential vocabulary words is included in the planning template for the predictive phase (Table 4.2). Throughout the lessons, students do a lot of discussing (talking and listening) and writing to ensure their understanding about the concepts. All of these tasks are associated with literacy.

Students can use the data collected (mass of the inflated balloon and the soccer balls) and find averages, construct graphs, and make predictions using the graphed data. All of these concepts are included in the National Council of Teachers of Mathematics standards (NCTM 2000) for this grade level.

Consideration Across the Grades

Students in the middle grades can be introduced to the idea of invisible particles and spacing of the particles and their movements. Display three charts with large circles to represent the position of the particles in each of the states of matter. Ask students to stand and form three groups. Each group arranges itself according to one of the charts (i.e., the students represent the particles in a particular phase). Include the movement of the particles in this investigation so that students can describe the movement of the particles as the matter goes from one phase to another. For example, when the matter is in the liquid phase, the particles are not as close together as in the solid phase, and as the matter evaporates, the particles are even farther apart. This activity continues to build the foundation for discussions that lead to the molecular structure of matter.

This lesson was designed for fifth-grade students; however, the lessons are also appropriate for use in third or fourth grade if your school teaches these concepts at these grades. If using the recommended balances exceeds the mathematical abilities of your students, consider weighing the items as a demonstration.

Build Your Library

These nonfiction trade books might be useful to help students understand the states of matter.

Solids, Liquids, and Gases (Ginger Garrett)
States of Matter: A Question and Answer Book (Fiona Bayrock)
What Is Matter? (Don L. Curry)
What Is the World Made of? All About Solids, Liquids, and Gases (Let's Read and Find…Science, Stage 2) (Kathleen Weidner Zoehfeld and Paul Meisel)
What's the Matter With Mr. Whisker's Room? (Michael Elsohn Ross)

Chapter 5

Earth's Shape and Gravity

CHAPTER 5: Earth's Shape and Gravity

Overview

As a graduate student in science education I sometimes attended the weekly symposium given by visiting educators. One of our visitors was Professor Yossi Nussbaum, an Israeli educator who shared my passion for teaching astronomy. As it happened, his presentation was a life-changing event.

Professor Nussbaum told us about efforts that he and Joseph Novak from Cornell University undertook to teach the spherical Earth concept to second graders in Jerusalem, Israel, and Ithaca, New York. Since the concept was commonly depicted in first-grade textbooks and there was a globe in nearly all classrooms, he imagined that most students already understood the basic idea. His goal was to help students understand the connection between the flat Earth of everyday experience and the spherical globe concept. The lesson—which was delivered as a sequence of slides with an audiotape—was cleverly designed to show students what Earth would look like if they were to leave Earth and fly into space.

However, the findings were very discouraging. Although the lesson was tested and revised several times, few second graders could grasp the concept. What was even more surprising was that when the students were interviewed after the lesson, most were perfectly satisfied that they understood why the apparently flat Earth is said to be round. Some of their explanations were ingenious: "Earth is really flat but people say it's round because of the hills." "Oh yes, we learned in school that Columbus proved the Earth is round. It's sort of like an island. He sailed from Spain and went all the way around the island and came back to the same port." Perhaps the most interesting was the student who said, "Yes, Earth is shaped like a ball. But we live on the flat part in the middle. The top part of the ball is the sky. The bottom part is the Earth. When the Sun goes under the ball, it can sometimes cause a volcano." One student even thought there were two Earths: the flat one we live on and the Earth up in the sky, "where the astronauts go" (Nussbaum and Novak 1976; Nussbaum 1979).

Later that day I had an opportunity to meet a group of elementary and middle school teachers, and I told them about the symposium. They assured me that all of their students learned about the spherical Earth concept in first grade, so they did not need to teach it at the upper elementary level. I invited the teachers to work with me to see if their students really did understand the concept, and three of them agreed to help. Over the next few months we interviewed 159 students in grades 3–8. To our surprise we found that Nussbaum was right. At the youngest ages very few students fully understood that we really live on a ball in space. Many of those who grasped the idea that Earth is ball-shaped thought that people must live just on top of the ball, because if people lived on the sides or underneath, they would fall off. Even among eighth graders, fully 25% of the students did not fully understand the spherical Earth concept.

With the assistance of a senior researcher, we published our findings in a refereed journal (Sneider and Pulos 1983). We also published an account of our research in a popular teacher's magazine, along with a questionnaire that teachers could use to find out what their students think about the Earth's shape and gravity and as a springboard for teaching the concept (Sneider et al. 1986). These findings have since been replicated and extended by many other researchers (Agan and Sneider 2004.) The research was also valuable in suggesting ideas for teaching about Earth's shape and gravity, and the original questionnaire used by the researchers has been refined by teachers over the years and now appears in this chapter as Figures 5.3–5.7.

Why This Topic?

It is not difficult to think of reasons why it is important for students to understand that a round globe represents our planet. After all, Earth is our home in space. Environmentalists remind us that it's the only home we have, so it's important that we take care of it. The spherical Earth concept is essential for understanding geography, weather, and certain phases of history, such as the great voyages of exploration in the 15th and 16th centuries. Learning about voyages to the Moon or space probes to other planets makes little sense if students fail to understand that Earth is one planet among several in the solar system. In other words, the spherical Earth and gravity concept is an essential building block for many other concepts that we expect students to learn about the world.

Yet the spherical Earth and gravity concept is among the most difficult for students to understand. Consider, first, the usual explanation for why Earth looks flat, even though we "know" it is round. Students are asked to imagine that they are ants crawling on a beach ball. The ant sees only a tiny part of the ball so, to the ant, the ball appears flat. If the ant could crawl all around the ball and come back to the same place, or sprout wings and fly up and away from the ball, it would perceive the spherical nature of its former home. Decades ago Piaget pointed out that the ability to see through another person's (or insect's) eyes—called perspective taking—is not easy for young children (Piaget and Inhelder 1956). It is an ability that comes with increasing maturity, and few second graders are able to see through the ant's eyes. Since the analogy is lost on them, the true nature of the globe remains a mystery to be solved in some other way.

Notice that I use the singular term *concept*, rather than *concepts*, when I refer to "the spherical Earth and gravity concept." Although in some elementary textbook series students are first introduced to the Earth's shape and then are introduced to gravity a year or two later, divorcing the two concepts is confusing. Most students immediately see the following problem: If people live all over the Earth, how come they don't fall off? Like adults, children don't tolerate conflicting ideas very well, so they think of ways to resolve the conflict between the ball-shaped globe and the obviously flat Earth beneath their feet. That is the origin of the many ingenious ideas that students invent to reconcile the two views—like the idea that "we live on the flat part in the middle," or the "two Earths" solution previously mentioned.

CHAPTER 5: Earth's Shape and Gravity

If we recognize children's need to make sense of their world, and respect their abilities to reconcile seemingly illogical ideas, then we must teach a rudimentary gravity concept at the same time we teach about Earth's spherical shape. Otherwise, we invite students to create their own understanding, which may be cute and creative, but at the same time erects barriers to later understanding. I say "rudimentary" gravity concept because Sir Isaac Newton's law of universal gravitation is both unnecessary and unnecessarily complex to introduce at this point.

In summary there are two reasons why it is difficult for students to understand the spherical Earth and gravity concept:

1. It is difficult to see through the eyes of an ant crawling on a beach ball, so it is hard to see why Earth is thought to be a sphere when it is obviously flat.
2. Students will immediately question the ball-shaped Earth concept when they try to envision people living all over the ball because they are accustomed to thinking about an absolute "down" direction in space.

If teaching the concept can be delayed until fourth or fifth grade, the first problem goes away because by then most students will have matured enough to see the beach ball through the ant's eyes and understand the analogy that is intended to reconcile the spherical and flat Earth models. Also, students will be old enough to understand a concept of gravity that is oriented toward Earth's center rather than toward an absolute "down" direction in space. However, these abilities will not be enough if we want students to understand the *evidence* and *logical arguments* that led people to accept the spherical Earth idea in the first place. Otherwise, understanding is simply an exercise in rote learning, rather than an engaging inquiry process.

If students cannot be expected to fully understand the Earth's spherical shape and gravity concepts until grades 4 or 5, teachers of early elementary grades may wonder why this chapter would be of value to them. First of all, the activities described in this chapter are *recommended* for students in grades 4 or 5 because that is the age at which most students can be expected to master these concepts. However, children develop at different rates, so many younger children may be able to successfully grapple with these ideas and find it challenging and enjoyable to do so. Although from a curriculum-planning perspective it is best to use precious science learning time to focus on lessons that the majority of students can master, the activities in this chapter can stimulate critical thinking and engaging class discussions throughout the elementary years.

A second reason why teachers of early elementary students may find this chapter useful is that it illustrates why students have difficulty conceptualizing certain ideas that involve complex visual thinking and highlights the kinds of cognitive abilities that students must develop during their first years of schooling to continue making progress in science. While some capabilities, such as perspective taking, develop naturally as children mature, teachers can help their students learn to apply these capabilities during daily classroom interactions, such as trying to understand a situation from another student's point of view. These capabilities lay the foundation for more complex thinking in the upper elementary grades.

To transform the task of teaching about Earth's shape and gravity from a seemingly intractable challenge into a great opportunity, we turn to the Instructional Planning Framework described in the early chapters of this book.

The Predictive Phase

In developing this chapter, I reviewed the process for the predictive phase and the hints outlined in Figure 2.2 (p. 24). I found the most helpful resources for articulating the concepts to be *Benchmarks for Science Literacy* (AAAS 1993), *Atlas of Science Literacy* (AAAS 2001b, especially pp. 42–43), and the *National Science Education Standards* (NRC 1996). For the purpose of unit design the most helpful books were *Understanding by Design* (Wiggins and McTighe 1998) and *Everyday Assessment in the Science Classroom* (Coffey and Atkin 2003).

Next, I sketched out the entire predictive phase and used the planning template (Table 5.1) to provide an overview. Filling out the planning template was especially helpful as it caused me to rethink some of my initial ideas for the lesson, so I went back and forth between the template and lesson development several times.

Table 5.1

Predictive Phase: Planning Template for "Earth's Shape and Gravity"

	Lesson Topic: "Earth's Shape and Gravity"
Essential Understandings	• The Earth we live on is shaped like a ball. It looks flat because we only see a small part of it. • People live all around the ball-shaped Earth and are held to Earth's surface by a pulling force called *gravity*. Gravity pulls everything toward Earth's center.
Knowledge Required From Previous Instruction	• A "model" is similar to but not exactly like the thing it represents (3–5). • "Earth" is a name for our entire world (3–5). • Gravity is a force that pulls things down, including us (3–5).
Knowledge and Skills to Be Learned	**Concepts:** See the Learning Targets. **Vocabulary:** *Earth, gravity, transparent, evidence, logic, phenomena* **Skills:** Ability to form and modify mental models; imagine the view from the perspective of another person (or animal); represent ideas visually; discuss ideas respectfully with other children; use simple logic; take evidence into account when making decisions

(continued)

CHAPTER 5: Earth's Shape and Gravity

Table 5.1 *(continued)*

Criteria to Demonstrate Understanding	<ul><li>Explain that Earth looks flat even though it is round because we see only a small part of it.</li><li>Respond appropriately to a "thought experiment" to imagine which way to look through a transparent Earth to (1) see people in countries far away or (2) envision how objects would fall when dropped in different locations or through tunnels dug into the Earth.</li><li>Use logic and the concept of gravity to explain why people who live in different parts of the world do not fall off.</li><li>Correctly predict that objects will fall toward Earth's center.</li><li>Formulate and argue a point of view by using evidence and logic. (Inquiry)</li><li>Consider two different mental models of the Earth to explain night and day and discuss which model is best and why. (Metacognition)</li></ul>
	**Learning Targets and Instructional Plans**
Target #1	*Model-building is a way of understanding our world.*
	Instructional Plan for Target #1: Will be completed during the responsive phase (see Table 5.2, p. 187).
Target #2	*A useful model for the Earth under our feet is that it is shaped like a huge ball in space.*
	Instructional Plan for Target #2: Will be completed during the responsive phase (see Table 5.2).
Target #3	*The Earth looks flat because we only see a small part of it.*
	Instructional Plan for Target #3: Will be completed during the responsive phase (see Table 5.2).
Target #4	*People live all around the ball-shaped Earth and are held to Earth's surface by a pulling force called* gravity.
	Instructional Plan for Target #4: Will be completed during the responsive phase (see Table 5.2).
Target #5	*Gravity pulls everything toward Earth's center.*
	Instructional Plan for Target #5: Will be completed during the responsive phase (see Table 5.2).

CHAPTER 5: Earth's Shape and Gravity

Stage I: Identify the Conceptual Target
(For a complete list of the ten stages for implementing the Instructional Planning Framework, see Table 2.1, p. 19.)

According to *Benchmarks for Science Literacy* (AAAS 1993, p. 63), in third through fifth grade all students should learn that "The Earth is one of several planets that orbit the Sun, and the Moon orbits around the Earth." However, the text also acknowledges that "In spite of its common depiction, the Sun-centered system seriously conflicts with common intuition. Students may need compelling reasons to really abandon their Earth-centered views. Unfortunately, some of the best reasons are subtle and make sense only at a fairly high level of sophistication" (pp. 62–63).

The *National Science Education Standards* (NRC 1996, p. 159) states that "By grades 5–8, students have a clear notion about gravity, the shape of the Earth, and the relative positions of the Earth, Sun, and Moon." However, the text acknowledges that although students in grades K–4 can learn to identify patterns of change in the sky, such as the movement of an object's shadow during the course of a day, "Attempting to extend this understanding into explanations using models will be limited by the inability of young children to understand that Earth is approximately spherical. They also have little understanding of gravity and usually have misconceptions about the properties of light that allow us to see objects such as the Moon" (p. 134).

In other words, all students should understand the Earth's shape and gravity concept by the time they reach middle school, but it won't be easy. Furthermore, choosing the right age level at which to present the concept is important. According to another pair of researchers who replicated Nussbaum's work in Nepal, "The danger is that the child will be told and will accept the new notions of Earth, gravity and space without understanding the meaning of the evidence or thinking about the implications of the new ideas" (Mali and Howe 1979, p. 685). Although it is true that students vary greatly in their developmental readiness to learn new ideas, the time when most students are prepared to understand and fully explore the implications of the Earth's shape and gravity concepts is in grades 4 or 5, when students "can begin to construct a model that explains the visual and physical relationships among Earth, Sun, Moon, and the Solar System" (NRC 1996, p. 159).

Stage II: Unpack the Standards and Identify the Concepts, Knowledge, Skills, and Vocabulary You Are Teaching

One of the most valuable outcomes of Novak and Nussbaum's seminal study was the development of a way to categorize students' alternative views of the Earth's shape and gravity concept. Their classification scheme is shown in Figure 5.1, page 178.

The five "notions" identified by Nussbaum and Novak (1976) can be envisioned as a sequence of mental models of the Earth's shape and gravity, ranging from a simple flat Earth model to the level of understanding anticipated by *Benchmarks for Science Literacy*

CHAPTER 5: Earth's Shape and Gravity

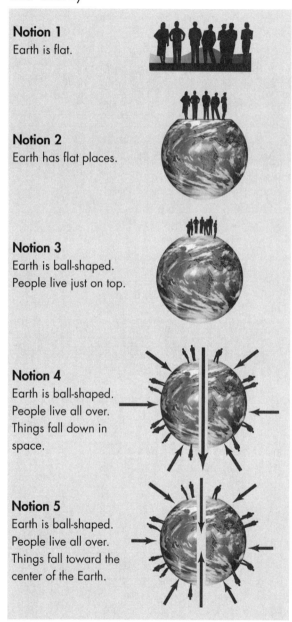

Figure 5.1

Students' Mental Models of Earth's Shape and Gravity

Notion 1
Earth is flat.

Notion 2
Earth has flat places.

Notion 3
Earth is ball-shaped. People live just on top.

Notion 4
Earth is ball-shaped. People live all over. Things fall down in space.

Notion 5
Earth is ball-shaped. People live all over. Things fall toward the center of the Earth.

Source: Illustration based on Nussbaum, J. 1979. Children's conceptions of the Earth as a cosmic body: A cross-age study. *Science Education* 63 (1): 83–93.

and the *National Science Education Standards*. The logical sequence of notions does not necessarily imply that all students progress from one step to another, since some students may skip some of the interim concepts. However, it does suggest a number of subconcepts, or *essential understandings,* that together make up the spherical Earth and gravity concept:

- The Earth we live on is shaped like a ball.
- The Earth looks flat because we only see a small part of it.
- People live all around the ball-shaped Earth and are held to Earth's surface by a pulling force called *gravity*.
- Gravity pulls everything toward Earth's center.

High school teachers will recognize that this concept of gravity is not the more advanced idea that every particle in the universe is attracted to every other particle. However, it is sufficient for students to understand the spherical Earth model as a logical whole.

Stage III: Identify the Learning Targets

Sputnik was launched when I was nearly 10 years old, and by then I already knew that I wanted to be an astronomer. I maintained that interest through my junior year in college when I discovered that teaching students was even more engaging and challenging than studying the stars. But by then I had enough experience with research in astronomy to realize that the process of astronomy is quite different from laboratory sciences such as physics and chemistry. Stars are simply too big and hot for hands-on experiments.

The name of the game for astronomers is model building. "Understanding" such phenomena as the birth and death of stars, the formation of planets and galaxies, or the nature of

CHAPTER 5: Earth's Shape and Gravity

quasars and black holes involves careful observation followed by the development of a conceptual or mathematical model of the phenomenon that would—if the model were correct—yield the same observations. An excellent example is spectroscopy. At college I had an opportunity to work in a spectroscopy lab where we observed the colors of light produced when subjecting various gases to high temperatures and pressures. If the gases glowed with the same colors as the light of a star, then we could be confident that we "understood" the composition and physical conditions in the star—even though it might be hundreds of light years away.

Students also build mental models of the world around them. Stella Vosniadou and William Brewer (1992, 1994) conducted a number of studies with first-, third-, and fifth-grade children, and found that their understanding of the day-night cycle depended on their mental models of the spherical Earth concept. For example, some students whose mental model of the Earth was a flat surface with an absolute down in space explained that day and night were caused by the Sun and Moon going up and down in the sky. In contrast, older students, who typically held a more advanced spherical Earth concept, tended to express misconceptions consistent with their mental model, such as the idea that the day-night cycle is due to the orbit of the Sun and Moon around a stationary spherical Earth. Such students have a much greater chance of correcting their misconceptions than students who still have a mental model of a flat Earth.

The point of this discussion is that if we want students to correctly explain astronomical phenomena, simply telling them about the phenomena is not sufficient. If they hold an incorrect mental model of the Earth in space, the new information may not make sense to them. So it is important for teachers to know how students envision the Earth as a whole, as a precursor to helping them correctly explain phenomena such as night and day, phases of the Moon, and so on. To develop a useful, mental model, students need to approach the spherical Earth concept as astronomers would—by constructing mental models to explain phenomena.

With this perspective we can now identify learning targets related to Earth's spherical shape and gravity:

- Learning Target #1: Model-building is a way of understanding our world.
- Learning Target #2: A useful model for the Earth under our feet is that it is shaped like a huge ball in space.
- Learning Target #3: The Earth looks flat because we only see a small part of it.
- Learning Target #4: People live all around the ball-shaped Earth and are held to Earth's surface by a pulling force called *gravity*.
- Learning Target #5: Gravity pulls everything toward Earth's center.

The connections among these learning targets are shown in Figure 5.2, page 180.

Aside from such simple terms as *up*, *down*, and *ball*, the only important vocabulary terms introduced in the lessons that will be described are *model*, *gravity*, and *Earth*.

CHAPTER 5: Earth's Shape and Gravity

Figure 5.2
Learning Targets for "Earth's Shape and Gravity"

Learning Target #1
Model-building is a way of understanding our world.

Model-building provides a foundation for considering the idea that...

Learning Target #2
A useful model for the Earth under our feet is that it is shaped like a huge ball in space.

This model only makes sense if students understand that...

Learning Target #3
The Earth looks flat because we only see a small part of it.

Students also need to recognize that...

Learning Target #4
People live all around the ball-shaped Earth and are held to Earth's surface by a pulling force called *gravity*.

Advanced understanding at the fourth- and fifth-grade levels is that...

Learning Target #5
Gravity pulls everything toward Earth's center.

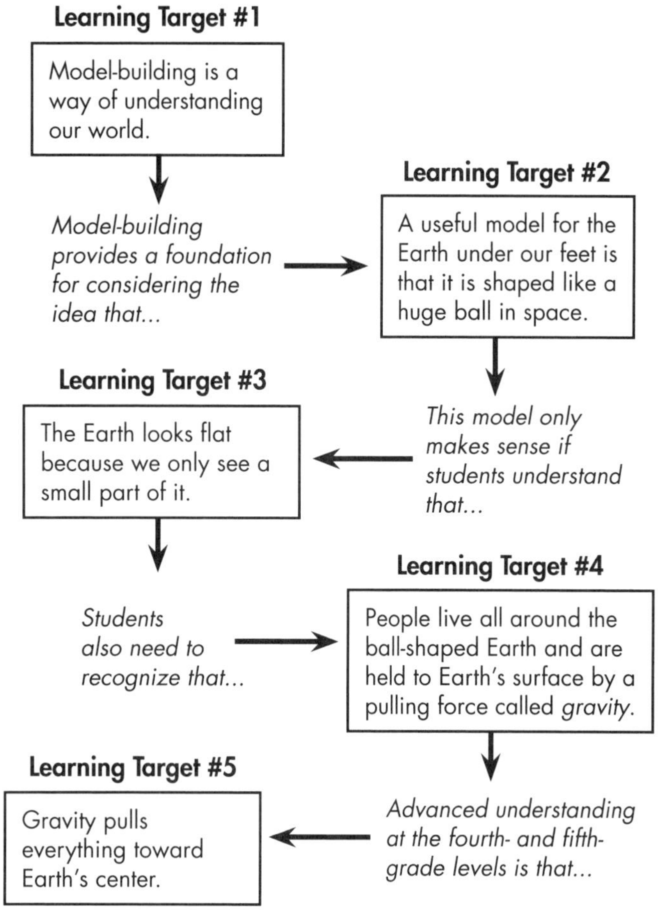

Prior to beginning a unit on the Earth's shape and gravity, it is worth spending some time discussing the concept of a model with the students, starting with models that they are familiar with, such as dolls and model cars. These models are like the thing they represent in some ways, but not other ways. The term *gravity* will be defined in context of the lesson, and developing a model for the Earth is the point of the entire lesson sequence. Students do not need to acquire any additional concepts before beginning this unit.

Stage IV: Identify Criteria for Determining Student Understanding

When I began working with a group of teachers to test Nussbaum and Novak's findings (1976), one of the first questions that occurred to us was *Why didn't anyone notice this problem before?* Since most textbooks taught the idea of a spherical Earth in first grade, it seemed odd that teachers had failed to notice their students did not understand the concept. The answer came from the students that we interviewed. They all said that the Earth is "round," suggesting that they understood the idea. Nussbaum and Novak (1976) were able to find out what the students really meant when they said "round" by asking them to apply the concept in various situations. Similarly, Vosniadou and Brewer (1992, 1994) were able to find out about students' mental models of the Earth by asking them to explain the day-night cycle. We asked similar questions and made the same discovery.

Except for Learning Target #1, which was developed to teach the metacognitive target ("Model-building is a way of understanding our world"), we used these research studies to develop ways to determine students' understanding of the learning targets. All of the activities in this chapter are embedded assessments that students carry out as part of the lesson. From the students' point of view, these activities are fun and thought-provoking tasks. From the teacher's viewpoint, the students' responses are indications of the students' current mental models.

CHAPTER 5: Earth's Shape and Gravity

Figure 5.3

Day and Night on a Flat Earth

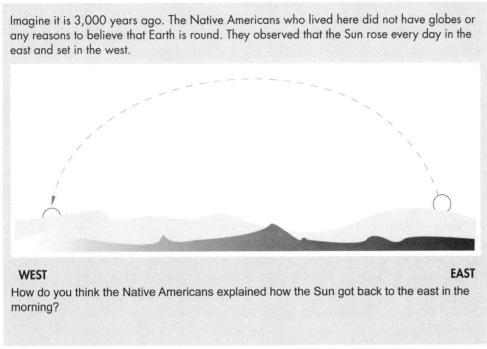

How do you think the Native Americans explained how the Sun got back to the east in the morning?

Note: Full-page pdfs of Figures 5.3–5.7 are available to download at *www.nsta.org/publications/press/extras/hardtoteachscience.aspx.*

Learning Target #1 (Model-building is a way of understanding our world.)

To find out if students are able to create a model that explains a phenomenon, teachers need to give them opportunities to construct at least two different models to explain the same phenomenon. As will be described in a subsequent section, students use both a flat Earth model and a spherical Earth model to explain the day-night cycle. The first task, shown in Figure 5.3, is a fun, creative activity that illustrates the students' logical use of a flat Earth model to explain how the Sun gets back to the eastern part of the sky after setting in the western part of the sky.

Learning Target #2 (A useful model for the Earth under our feet is that it is shaped like a huge ball in space.)

It's easy for many students to envision a huge ball in space, until they realize that the "ball" that the teacher is talking about is the very Earth they are standing on! After all, it seems preposterous that this solid Earth is floating with no support and hurtling through space on a vast journey around the Sun. The criterion for determining if

CHAPTER 5: Earth's Shape and Gravity

Figure 5.4
Imagining People on the Other Side of the Earth

1. Imagine that you have x-ray vision, and you can look all the way through the Earth as though it were made of glass. Which direction would you look to see people in far-off countries like India and Australia?

    A. Eastward

    B. Westward

    C. Northward

    D. Southward

    E. Upward

    F. Downward

2. When you look through the Earth and see people far away, which part of their bodies do you see?

    A. The tops of their heads.

    B. Their faces.

    C. The bottoms of their shoes.

    D. The backs of their heads.

    E. Their profiles (side view).

See note for Figure 5.3, page 181.

students realize that the lesson is about the Earth under their feet is for them to answer a question in which they must apply the spherical Earth concept to the ground they are standing on.

To find out if students really grasp this idea, ask students to imagine that they have superpowers and can look right through the Earth to see people in far-off countries. Students who understand this concept will realize that they would not look eastward or southward, as a plane might fly to those countries, but look downward instead. A second question, concerning which part of a person's body they would see when they looked right through the Earth (bottoms of their shoes), provides an additional check on their mental model. These questions are shown in Figure 5.4

Learning Target #3 (The Earth looks flat because we only see a small part of it.)

For students to make sense of the spherical Earth concept, they need to understand the ant-on-the-beach-ball analogy: Earth appears flat to us because we only see a small

Figure 5.5

Is the Earth Round or Flat?

In school we learn that the globe represents the round Earth. But if we look outdoors, the Earth looks flat. Why is that?

A. They are different Earths. We live on the flat Earth. The round Earth is a planet in the sky.

B. The Earth as a whole is round. We live on the flat part in the middle.

C. The Earth is round like an island. That's why it can be flat and round at the same time.

D. The Earth is round like a ball. It only looks flat because we see just a tiny part of it.

See note for Figure 5.3, page 181.

part of its surface. Suggest to students several alternative explanations that represent common misconceptions to see if the students are able to distinguish which is the correct explanation. Figure 5.5 presents such a task. The students are presented with two images of the Earth: an Earth globe and a picture showing a flat horizon. They are asked to choose from a list of possible explanations for why Earth appears as a ball in one image and as a flat surface in the other image. Students who choose answer "D" have a scientific model of the Earth—they are able to simultaneously take the viewpoint of the ant crawling on a ball *and* of a person looking at the ant on the ball.

Learning Target #4 (People live all around the ball-shaped Earth and are held to Earth's surface by a pulling force called gravity.)

To understand how people can live all over the Earth without falling off, students need to have a rudimentary understanding of gravity: a force that holds people to Earth's surface. And to understand that idea, they must give up their idea of an absolute "down" direction in space. The criterion for determining if students grasp this idea is

CHAPTER 5: Earth's Shape and Gravity

Figure 5.6

Which Way Will the Apple Fall?

The picture below shows a drawing of five cooks living in different parts of the world, each dropping an apple. For each apple, draw the path it will follow as it falls. Draw a dot at the end of the line showing where the apple lands.

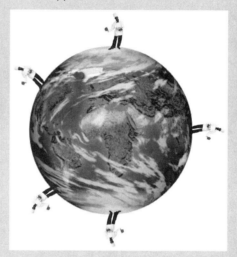

Explain why the apples will fall this way.

See note for Figure 5.3, page 181.

to have them respond to a drawing in which they must choose between an absolute *down* direction in space versus *down* toward Earth's surface.

A question that provides evidence of whether or not students grasp the idea of gravity is shown in Figure 5.6, in which students are asked to imagine people all around the Earth, draw what would happen to an apple dropped at various places, and explain their answers. Evidence for full understanding would include lines from the apple in each figure pointing directly toward Earth's surface and an explanation about how gravity pulls things down toward the Earth.

Learning Target #5 (Gravity pulls everything toward Earth's center.)

A criterion to determine if students understand this concept is that they can ignore both the idea that there is an absolute "down" direction in space and that gravity acts just at the surface of the Earth. Instead students should recognize that even inside Earth, gravity pulls things toward its center. Figure 5.7 is a "thought experiment" that shows an imaginary hole drilled all the way through Earth from pole to pole and asks students to

Figure 5.7

Exploring a Tunnel Drilled Through the Earth

Below is a diagram, showing an imaginary tunnel drilled all the way through the Earth from pole to pole. If a cook drops an apple into the hole, draw the path of the apple. Make a large dot at the end of the path showing where it lands.

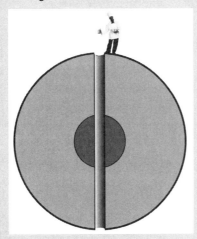

Explain why the apple follows this path.

See note for Figure 5.3, page 181.

imagine what would happen if an apple were dropped into the hole. Students who have met this learning target will draw the apple falling into the hole and eventually stopping in the center. A physics teacher would recognize that the apple would not stop in the center, but would pass through the center and fall back and forth, slowing down because of friction with air, and eventually settling in the center. However, it's also okay for this grade level if students draw the apple falling directly to the center and stopping there, as it indicates they understand that gravity pulls objects toward Earth's center.

There is an inquiry element to all five learning targets, in that they are not to be learned by rote, but rather through an inquiry process, so that students have opportunities to deconstruct any misconceptions they may have and construct a more adequate understanding of science and of the Earth's shape and gravity. Consequently, this last criterion is for students to formulate and argue a point of view by using evidence and logic. This capability cannot be measured by test items but only by observing and listening to students as they debate their various ideas. As will be explained in subsequent sections, there are many opportunities to do that during the proposed seven lessons.

CHAPTER 5: Earth's Shape and Gravity

Stage V: Inquiry and Metacognitive Goals and Strategies

Metacognitive Goals

According to the National Research Council's report on findings from educational research (Duschl, Schweingruber, and Shouse 2007; Michaels, Shouse, and Schweingruber 2008), there should be two types of metacognitive goals in science education: (1) learning about the nature of science and (2) engaging students in reflecting on their own scientific thinking. The first type of goal corresponds to Learning Target #1—that astronomy is not an experimental science but rather a process of building mental models that explain phenomena. It is important to engage the students in building mental models—such as the model of Earth discussed in this chapter—and to have them become explicitly aware that constructing mental models is one way of doing science. The second type of metacognitive goal is to engage students in reflecting on how their thinking has changed at strategic points during the lesson.

Creative thinking and learning is the strategy that I selected for introducing students to what astronomers do. As elaborated in Table 2.4 (p. 31), this strategy involves "generating new ways of viewing a situation outside the boundaries of standard conventions." It is a fun, creative activity in which the students explain night and day from a flat Earth point of view. Specific strategies for getting the students to visualize and articulate their invented models include drawing and speaking about their ideas in front of the class. The teacher then uses this experience to illustrate that mythological thinking was a precursor of the science of astronomy, which also builds models to explain phenomena. The difference between mythology and the science of astronomy is that astronomy has the added constraint of seeking a deeper truth based on evidence. The strategy then switches to a different metacognitive strategy—critical thinking—in which the students fully explore the implications of the spherical Earth and gravity model, applying both evidence and logic.

Inquiry Goals

Parallel to the metacognitive strategy for introducing the first learning target—creative thinking—the corresponding inquiry strategy is open-ended inquiry. There is no single right answer to the question, "If you believe Earth is flat, how do you explain how the Sun gets back to the east every morning?" Students are encouraged to draw on what they already know about Native Americans or simply imagine what people may have thought long ago (see Figure 5.3). A parallel shift in inquiry then takes place. As the teacher shifts the metacognitive goal to critical thinking and learning when introducing Learning Targets #2, 3, 4, and 5, the inquiry strategy shifts from open-ended inquiry to guided inquiry.

I should emphasize that the inquiry process should be lightly facilitated. Although there are right and wrong answers to the questions posed in Figures 5.4, 5.5, 5.6, and 5.7, it is essential that the students come to that understanding by interacting with their peers, through small-group and large-group discussions. The teacher should present the modern scientific perspective only after the students have thoroughly explored the issues. Manipulatives (Earth globes removed from their stands) are also very helpful aids to inquiry thinking as is the use of drawings to draw out students' thinking.

This rounds out the predictive phase of lesson development. We have identified the learning targets, established inquiry and metacognitive goals, and know how we will find out whether or not students have reached these targets and goals. Now we turn to planning a sequence of lessons.

The Responsive Phase

Once again I found the process outlined in Chapter 2 helpful in developing these lessons. The process for the responsive phase and the various hints and strategies were reviewed and used in this chapter's development. The following sections describe the results of that process, and a summary of the results for the responsive phase is found in the planning template for the responsive phase of the Earth's shape and gravity lessons (Table 5.2).

Table 5.2

Responsive Phase: Planning Template for "Earth's Shape and Gravity"

Identify Student Preconceptions	The preliminary activities associated with each lesson will provide information about your students' mental models of the Earth's shape and gravity. You can use their responses to group students for instruction or decide which lessons can be shortened or skipped.

Learning Target #1
Model-building is a way of understanding our world.

Instructional Plan for Learning Target #1
(This plan does not account for changes that will come about during instruction or re-teaching.)

Engagement

The purpose of this first activity is to engage students in the scientific process that is peculiar to astronomy and a few other fields—that is, to develop a model that explains a phenomenon.

1. *Ask:* Imagine that we are Native Americans who lived thousands of years ago. We notice that the Sun sets in the west and rises in the east every day. If we believed the Earth is flat, how might we explain how the Sun goes from the west back to the east during the night?
2. Students work in pairs to create possible explanations and draw their ideas. Provide each pair with a copy of Figure 5.3.
3. Students report on their ideas to the whole class.
4. Validate students' creative models by explaining that such stories formed the roots of science. The model of a ball-shaped Earth arose in ancient Greece, at which time people began to think of models that were supported by evidence and logic.

(continued)

CHAPTER 5: Earth's Shape and Gravity

Table 5.2 *(continued)*

Learning Target #2: *A useful model for the Earth under our feet is that it is shaped like a huge ball in space.*
Learning Target #3: *The Earth looks flat because we only see a small part of it.*
Learning Target #4: *People live all around the ball-shaped Earth and are held to Earth's surface by a pulling force called gravity.*
Learning Target #5: *Gravity pulls everything toward Earth's center.*

As described in the next section, teaching each of these targets one at a time involves a similar sequence.

Instructional Plan for Learning Targets #2, 3, 4, and 5
(This plan does not account for changes that will come about during instruction or re-teaching.)

Elicit Students' Preconceptions

1. Begin by having students write their answers to a question. For example, when starting a class aimed at Target #2, have students write their individual answers to the question depicted in Figure 5.4, which asks which way you would look to see people in far-off countries such as India and Australia. (Targets #3, 4, and 5 begin by having students answer questions posed in Figures 5.5, 5.6, and 5.7.)

Confront Preconceptions: Engage Students in Small- and Large-Group Discussions

2. Form small groups of three to four students. Give each group one copy of the same figure and ask them to come to agreement on the best answer. Provide an Earth globe that is not in a stand to help groups figure out the best answer.
3. Lead a large-group discussion in which all of the proposed answers are considered and debated.
4. Facilitate discussions to be sure that several different ideas are discussed, and encourage students to consider both evidence and logic.

Sense Making: Encourage Discussion Before Providing Scientific Answers

5. Do not provide the scientific answers to the questions until the students have had time to talk among themselves. If possible, delay giving the scientific answers until the next day and encourage students to continue the discussion among themselves or with parents and siblings that night.
6. Provide scientific explanations in response to the questions, taking time to comment on good ideas that students have suggested. Emphasize that these modern scientific ideas are models developed to explain natural phenomena. These models have since been confirmed by evidence (such as trips around Earth by astronauts).

Extend and Elaborate

7. Have students use the internet to research how members of the Flat Earth Society explain phenomena such as the day-night cycle, paying attention to the society's members' model of the Earth and noting how it is supported by evidence and logic.

Stage VI: Research Children's Misconceptions Common to This Topic That Are Documented in the Research Literature

The many variations of student misconceptions about the Earth's shape and gravity can be sorted into the following three major categories.

- *Flat Earth Models.* The five "notions" described by Nussbaum (1979) shown in Figure 5.1 provide a good overview of the common misconceptions that students develop about Earth's shape and gravity. Because students perceive the Earth to be flat, they have great difficulty understanding a lesson in which they are told Earth is a sphere. This is especially true if students are unable to understand the ant-on-a-beach-ball analogy. Many students invent creative ways to make sense of the spherical Earth concept while not giving up their flat Earth conceptions. The ideas that the ball-shaped Earth refers to hills or to a planet in the sky "where astronauts go" fit in this category (Vosniadou and Brewer 1992, 1994).
- *Ball-Shaped Earth Models.* Once students understand that the spherical globe is intended to represent the Earth beneath their feet but still conceive of an absolute down direction in space, they may develop mental models that accommodate both ideas, such as "Earth is shaped like a ball and people live on top of the ball" or "We live on the flat part in the middle." These two models are representative of this category (Agan and Sneider 2004; Sneider and Pulos 1983).
- *Gravity Models.* Many students in upper elementary grades understand that the Earth beneath their feet is a huge sphere and that people live all around the sphere and are held to the surface by gravity. They understand that from the viewpoint of any individual on Earth, they are right-side up, and everyone else is upside-down. However, many of these students think of gravity as a pushing force due to air pressure or some other force that holds people to the surface and are confused when asked to think about gravity in tunnels inside the Earth (Agan and Sneider 2004; Bar, Sneider, and Martimbeau 1997).

Stages VII–X all involve strategy selection, beginning with the use of the Instructional Tool 2.3 (p. 42) and then accessing information in the various other instructional tools. Table 5.3, page 190, provides an overview of strategy selection for this unit and the next sections of the chapter further detail use of the strategies.

Stage VII: Select Strategies to Identify Students' Preconceptions

Take care to ensure that the students respond in writing to the various questions in Figures 5.3, 5.4, 5.5, 5.6, and 5.7 so that it will be easier to interpret their drawings later. You could present just one of the drawings at a time, or two or three at once. Reading over the students' papers will provide you with a clear picture of your students' mental models of the Earth's shape and gravity. If some of your students have flat Earth

CHAPTER 5: Earth's Shape and Gravity

Table 5.3

Strategy Selection for "Earth's Shape and Gravity"

	Learning Target #1*	**Learning Target #2**	**Learning Target #3**	**Learning Target #4**	**Learning Target #5**
	Model building is a way of understanding our world.	A useful model for the Earth under our feet is that it is shaped like a huge ball in space.	The Earth looks flat because we only see a small part of it.	People live all around the ball-shaped Earth and are held to Earth's surface by a pulling force called *gravity*.	Gravity pulls everything toward Earth's center.
Possible Strategies for:					
Identifying Preconceptions	Specific strategies to identify student preconceptions were not used. Instead, the development of the unit was based on student misconceptions that had already been identified in the research.				
Eliciting and Confronting Preconceptions		• Questioning and discourse • Small- and large-group discussion	• Questioning and discourse • Small- and large-group discussion	• Drawing out students' thinking • Small- and large-group discussion	• Drawing out students' thinking • Small- and large-group discussion
Sense Making	Model-building	• Model-building • Manipulatives	• Model-building • Manipulatives	• Model-building • Drawing	• Model-building • Drawing
Demonstrating Understanding	Visual models and speaking	Visual models and discourse	Visual models and discourse	Visual models and discourse	Visual models and discourse
Selected Inquiry Strategy	Open-ended inquiry	Students communicate and use logic to justify why one mental model provides a better explanation for a phenomenon than another model.			
Selected Metacognitive Strategy	Creative thinking and learning	Critical thinking and learning			

* The goal of the first learning target is to establish a pattern of thinking that parallels the process of science in astronomy—to build models for explaining natural phenomena. This is a creative activity, so it is not necessary to draw out students' preconceptions at this point.

conceptions, it's best to start at the beginning so they have a chance to listen to the other students and catch up. However, if all of your students understand the Earth is shaped like a ball, then it may be best to begin with their ideas about gravity.

You will note that the instructions in each of the figures ask students to also explain their thinking in writing. It is important to encourage your students to write, as writing is a very helpful way for the students to sort out their own thinking, as well as for you to find out what they mean by their drawings. This idea is expressed in Instructional Tool 2.9 (p. 121), which notes that "A combination of visual and verbal strategies…lets us look for a match between our students' verbal and visual representations and lets our students elaborate on their understandings." So we should use drawings and have students discuss their drawings to better address misconceptions.

Stage VIII: Select Strategies to Elicit and Confront Your Students' Preconceptions

Once you know how the different students in your class think about the Earth's shape and gravity, you can form small groups of three or four students with a range of views. By assigning each group the task of agreeing on a single response to the same question, you will initiate a process in which students will articulate their preconceptions and find that not everyone agrees with them. If the confrontations come from peers, the students will be much more likely to engage in lively conversation than if you confront their ideas as an adult authority figure.

This strategy is described in Instructional Tool 2.6 (p. 103), which states, "Better discussions occur when students see several plausible viewpoints, resulting in students' own theory determination. When only one viewpoint is presented, there is no reason to have a discussion."

Stage IX: Select Sense-Making Strategies

Providing physical objects to small groups of students to model their ideas is a very helpful strategy for many students. Students can use inflatable Earth globes without stands to demonstrate that any of Earth's continents can be placed "on top." Some Earth globes are partially transparent so it is possible for students to see that they would have to look downward through a transparent Earth to see people in countries far away.

The use of all types of models is encouraged in Instructional Tool 2.7 (p. 107), which notes that "Generating a concrete representation creates an 'image' of the knowledge in a student's mind. Physical models, which are one form of concrete representations, enhance nonlinguistic representations and students' understanding of content."

While the globes are useful, the most important models used in this series of activities are the drawings in Figures 5.3–5.7, which engage your students in visualizing the Earth as a huge ball in space. This idea is described in Instructional Tool 2.7 (p. 112), as follows: "Graphic models can be used to determine students' preconceptions. Presenting a graph, table, or figure and asking students to describe/interpret the image allows you to determine what they know and/or misunderstand about the represented ideas. This can be done individually or in small groups. If you have your students share and critique their developed models, it will help them see connections between ideas and help you recognize gaps and changes in student understanding."

The previous paragraph refers to the value of graphic images in eliciting students' preconceptions. Drawings are also a very powerful means for students to make sense of conflicting ideas because drawings allow them to visualize alternative arguments and find the holes in their own thinking processes.

Another useful strategy is to pull together small groups into a large-group discussion and project, write, or draw on the board ideas from the different groups so students can debate the top three or four ideas in response to each of the questions in Figures 5.3–5.7. You should not tell students what your own choice is until after they have discussed their ideas and taken a vote on the most likely answers. In fact, it's a good idea to leave your own analysis until the next day, so the students have a chance to talk about their ideas after class or at home with their parents.

Stage X: Determine Formative Assessments

When making copies of the questions in Figures 5.3–5.7, it's a good idea to make *two or three copies for each student* in your class, so you can have each student respond at the start of the unit and at least one more time after some of the discussions to see how his or her ideas are changing as a result of discussions with other students. Save the students' papers from the beginning of the unit. At the end of the unit, present the same questions again. When the students have finished drawing and writing their final answers, hand out the papers they completed before the unit so they can see how their own thinking has changed. Ask them to discuss what convinced them to change their minds. As mentioned in Instructional Tool 2.2 (p. 83), "We should constantly encourage students to compare their current thinking to their original thinking and try to determine what helped them achieve their current understandings." You might consider providing a writing frame (Fulwiler 2007) such as "I used to think _____ because ____, but now I think _____ because _____."

It is also revealing to listen carefully to the discussions. Some teachers like to appoint a student to lead a whole-class discussion on each of the questions, so they can sit on the sidelines and listen carefully, noting expressions of students who are not actively participating and occasionally making suggestions to enrich the discussion.

If there is time after the discussion, ask the students how it felt to change their minds about certain ideas. Sometimes the "aha!" can be exhilarating, and sometimes it is uncomfortable to realize that a prior understanding was incorrect. Simply being aware of our emotions as we learn can be an important aid to further learning.

CHAPTER 5: Earth's Shape and Gravity

The Lessons: Teaching and Learning About "Earth's Shape and Gravity"

The following sequence of lessons grew out of the research studies discussed at the start of this chapter and were revised during pilot studies and with contributions from many teachers and curriculum developers. These ideas were eventually published as the *Earth, Moon, and Stars Teacher Guide* in the series Great Explorations in Math and Science (GEMS), published by the Lawrence Hall of Science (Sneider 1986). Some of these activities have since been revised and updated as part of an in-depth astronomy curriculum for grades 3–6 (Beals and Willard 2007).

Lesson 1: What Do Astronomers Do?

The first task, depicted in Figure 5.3, asks students to imagine that they are Native Americans, living several thousand years ago in the same region where your students live today. Like most people in the world at that time, Native Americans probably believed that Earth is more or less flat. It is also likely that they observed the Sun, Moon, and stars, and asked themselves questions about what they saw. For example, they observed that the Sun rose in the eastern part of the sky every day and set in the western part of the sky. They must have wondered how the Sun got back to the east again the next morning.

After being introduced to the task, the students are organized into pairs to find a creative answer to that question from the Native Americans' point of view and to draw their ideas with colored markers on a large sheet of paper. The next day, students put up their drawings on the walls and share their stories. Commonly, students will envision that animal spirits carry the Sun back to the east at night, possibly along a river or underground. Some students suggest that the Sun is quenched when it falls into the water at night, and a new Sun is born the next day.

After the students have shared their stories, assure them that their stories are much like the many myths told by people all over the world since ancient times. Myths such as these that create a vision or model of the world and how it functions were the roots of science and the beginnings of scientific theories.

Explain that the spherical Earth idea was first proposed in ancient Greece, where travelers from all over the world shared their ideas about the shape of the world. Some of the Greeks thought that there might be a way of thinking about the shape of the world based on logic that people could agree on rather than just assuming that the stories they had been told were true.

The importance of this first lesson is twofold. First, it communicates the nature of astronomy as a process of observing a natural phenomenon (in this case, the disappearance of the Sun at night and its reappearance on the other side of the sky in the morning) and formulating a model that might explain the phenomenon. Second, it provides an opportunity for students who still hold a flat Earth model to engage in the

CHAPTER 5: Earth's Shape and Gravity

activity at the same level as all the other students and confront a phenomenon that is not so easy to explain with a flat Earth model.

Lesson 2: Where Do People Live?

Start by handing out a copy of Figure 5.4 to each student and asking the students to circle the answers to the two questions. This task asks students to imagine that Earth is transparent and to say which way they would look to see people in far-off countries. Have students record their answers a few minutes before the end of one period, so you can look at the students' papers overnight. Separate the papers of the students who answer correctly (1F and 2C). If all the students answered correctly you can skip this lesson; if several gave other answers, however, it is worth spending a class period on it.

For Lesson 2, organize the class into groups of three to four students. If possible include at least one student in each group who answered both questions correctly. Give each group one blank copy of Figure 5.4 and ask them to come to agreement on the best answer. If possible provide each group with an inflatable Earth globe. Allow 15–20 minutes of discussion, or until each group has agreed on the best answer.

Call the whole class together and allow students from each group to report and explain why they answered as they did. If there are disagreements, take a vote, then allow more discussion and another vote to give students a chance to change their minds. Listen respectfully rather than providing the right answers. If the students still disagree at the end of the period, continue the next day by asking if anyone changed his or her mind overnight. Then explain how you would answer the question, while validating the reasonable and creative ideas given by students who may still disagree. Acknowledge that it's really hard to believe that people live down beneath our feet!

In some ways this may be the most difficult of the tasks that your students will face as they try to reconcile what they see with what they are learning about the world as a planet in space. Few people ever stop to think that people actually live down below their feet, on the other side of the world. Kids who ask if it's possible to "dig all the way to China" fully grasp this idea.

Through this activity, students understand that discussion about the Earth, as a ball in space, is not something "out there" that does not affect them. It requires that they look at the Earth of their everyday experience. If any of your students experience an "aha!" event and realize that the Earth they are standing on is actually a huge ball, they will truly be welcomed as a resident of Planet Earth.

Lesson 3: Why Does Earth Look Flat?

Figure 5.5 is the focus of the third lesson. Hand out a copy of Figure 5.5 to each student and ask him or her to circle the correct answer. Separate the papers of the students who answer correctly (D). If all the students answer correctly, skip this lesson; however, if several give other answers, then organize the class into small groups of students who gave diverse answers and have each group decide on the best answer.

As in the prior lesson, allow 15–20 minutes for small-group discussions, using more or less time depending on how the discussions are going. Hold a large-group discussion and then take a vote or two; encourage students to argue their positions without giving away your position. If there is still controversy, wait until the next day to offer your answer. If students have not brought it up, this is a good time to introduce the ant-on-a-beach-ball analogy.

Through this activity students reconcile the spherical Earth model with the flat Earth of their daily experience, which is absolutely essential if they are to fully adopt the model as their own.

Lesson 4: Why Don't People at the "Bottom" of the Earth Fall Off?

Even if students have struggled with the spherical Earth model, move on to Figure 5.6 because some students will need to think differently about gravity to have the spherical model make sense. As before, have students individually record their answers. When you distribute Figure 5.6, tell the students to put their pencil points on the apple and draw the entire path of the apple when the cook lets it go. They should draw a dot at the end of the line showing where the apple lands. They should do the same for each of the five apples, then write a sentence explaining why they drew the paths as they did.

As before, review the students' papers and decide how to proceed. Organize small groups with students who have diverse ideas, allow time for them to discuss in their groups, and then conduct a large-group discussion. Facilitate the discussion by projecting or drawing three or four representations on the board and inviting students to show their ideas to the rest of the group. Invite discussion and take a vote, followed by more discussion and a final vote.

If there has been a good discussion, this is an excellent time to ask students to reflect on how their thinking has changed: What arguments did you find most convincing? How did it feel to change your mind? Does the ball-shaped Earth model make more sense now?

Understanding the spherical Earth concept at this level can be considered proficient for a fourth- or fifth-grade student. That is, a solid grasp of the spherical Earth concept involves understanding that

- people live all around a ball-shaped Earth;
- if Earth were transparent, we could (theoretically at least) see people in far-off countries living down beneath our feet;
- Earth looks flat because we only see a small part of it; and
- gravity is a force that holds us to Earth's surface.

CHAPTER 5: Earth's Shape and Gravity

With these understandings, your students can go on to study phases of the Moon, space probes to other planets, and so on.

However, some students will be able to go beyond the proficient level and will learn about gravity as a pulling force that acts even inside the Earth to pull things toward Earth's center. So consider the next lesson a "stretch goal" to challenge the students who are ready to expand their mental model of gravity. All students should have a good time thinking about this question—"What happens to objects that fall inside the Earth?"—even if they are not able to advance to the next level. The most advanced students should also enjoy the discussion; it has even caused lively discussion among college physics professors!

Lesson 5: What Happens to Objects That Fall Inside the Earth?

The general approach is the same as in the prior lessons. When introducing the task in Figure 5.7, explain that the drawing shows an imaginary hole drilled through the Earth from pole to pole. Your students may challenge the idea that such a hole could possibly be dug because of the great temperatures and pressures inside the Earth. Of course they are perfectly correct! Explain that this is a "thought experiment" to see how they think gravity would act inside the Earth. As in Figure 5.6, they should put their pencil points on the apple and draw the entire path of the apple, with a dot showing where it will stop.

Collect their papers, look at them overnight, and organize small- and large-group discussions as in prior lessons. Usually the class ends with some students expecting that the apple will go to the center of the Earth and stop, while other students think the apple will pass the center, and then "fall back" toward the center, possibly falling back and forth a few times before stopping. But there are other possible answers as well. Some students have thought that the apple will go all the way through the Earth and "go into orbit" or that it will stick to the sides of the tunnel. One student claimed the apple would refuse to go into the hole at all!

I have always found that students really want to know, "What will really happen to the apple?" That is why I like to string it out for at least a day and urge the students to ask their parents what they think will happen to the apple. (Give the students an extra copy of Figure 5.7 to take home with them.) When I do provide the scientific answer to the question, I emphasize that scientists have spent a lot of time thinking about this question—just as the students themselves have (see Figure 5.8 for examples of the thinking of two famous scientists).

A common misconception, shared by many children and adults today, is that objects will fall to the center of the Earth and stop because that is "the center of gravity." In fact, the center of Earth—or any other body—has no special gravitational attraction. That term is used to describe a balance point, around which all mass is equally distributed. According to Newton's theory of gravity, an object placed in a tunnel at Earth's center would not move at all, since all gravitational forces on it would be equal.

Figure 5.8

Scientists Through the Ages Have Grappled With Ideas

> Aristotle was a Greek philosopher who lived more than 2,000 years ago. Based on his writings we can be sure that he would have predicted that the apple would fall to the center and stop. Evidence and logic convinced him that Earth is shaped like a ball. For example, he noted that the shape of Earth's shadow on the Moon during a lunar eclipse is always curved; a journey to the south reveals certain stars in the night sky that cannot be seen at all farther to the north; and different stars pass directly overhead in different north-south locations (Aristotle c. 350 BC/1971). Later, other philosophers noted that when a ship was first sighted its sails were always spotted first, and then the hull would come into view as it approached.
>
> Aristotle reasoned that if Earth were shaped like a ball there must be an explanation for why people who live all around the ball don't fall off. He reasoned that Earth must be the center of the universe and that everything that has weight would fall to its center. In other words, he would have predicted that the apple would fall to the center of the Earth and stop because it reached the center of the universe.
>
> About 2,000 years later, in the 17th century, Sir Isaac Newton had a different idea—that gravity is a pulling force between every particle of matter in the universe (Newton 1687/1972). He would say the apple would fall into the hole and go past the center. Once it is past the center, more Earth particles would be pulling it back to the center. So, after it passes the center, it would slow down, stop, and "fall back" toward the center. If there were no air in the hole, the apple would fall back and forth forever. If the hole were filled with air, the apple would gradually slow down and settle in the center of the Earth.

However, when an object is dropped into such a tunnel from Earth's surface, it would continuously speed up as long as there is more matter ahead of its fall than behind it. By the time it reaches the center, it would be moving very fast, so it would not stop. It would just start to slow down after passing the center.*

It is not essential for all students to fully understand what will happen to the apple, at least not until they study Isaac Newton's theory of gravity in high school physics. When they do study physics, they will have a leg up on students who are challenged by developing an intuitive understanding of those more counterintuitive ideas.

Lesson 6: What Do You Say to a Member of the Flat Earth Society?

The Flat Earth Society is an organization of people who believe that the Earth is not shaped like a ball at all. They believe it is just as it appears—flat. The organization began in the 19th century and thrived throughout most of the 20th century. The society has a website and posts a number of interesting articles that anyone can access at *http://theflatearthsociety.org/cms*.

---

*High school physics teachers will recognize that air friction will slow the apple down until it reaches "terminal velocity," which will occur long before the apple reaches Earth's center.

CHAPTER 5: Earth's Shape and Gravity

What is most interesting about the articles, debates, and other posts on this site are the various mental models that are proposed if one believes that Earth is indeed flat. For example, does the flat Earth's surface go on forever? Or is it as some students have imagined, a round island with a circular edge? How far does it go downward? Is it a flat slab with a bottom, or is it infinite in depth? If the Earth is flat, then how does the Sun get from the western sky back to the eastern sky in the morning?

This last question—How does the Sun get from the western sky back to the eastern sky in the morning?—provides a final check on the students' understanding of the spherical Earth concept; ask students to tell you their current understanding of how the Sun gets back to the eastern sky in the morning. If they reply that it's not the Sun that moves, but rather the Earth that turns, making it *appear* that the Sun went around to the other side of the sky, then they have a good grasp of the spherical Earth concept. Then have the students access the Flat Earth Society's website to find out how its members explain the day-night cycle and read some of the other articles and posts. A good follow-up assignment is to have the students write a few paragraphs about an imaginary conversation with a member of the Flat Earth Society about what causes day and night. What would that person say? How would that person defend his or her beliefs? How would the students in your class reply?

Alternatively, you could stage a mock debate. Have students draw from a hat to see if they are on the Flat Earth Team or the Ball-Shaped Earth Team. Divide teams into smaller groups of three to four students to prepare their arguments. Then have the teams take turns making initial statements, rebuttals, and cross-rebuttals. By thinking through these arguments, students will develop a much better understanding of the spherical Earth concept and the evidence and logic that supports it.

Lesson 7: How Have Your Ideas Changed?

Hand out a set of Figures 5.3–5.7 to all students and ask them to answer the questions one more time. This time they should add a sentence or two at the bottom of each sheet to say if their ideas have changed, and if so how. This can be done in class or as a homework assignment. It's best if the students are told that they will not be graded on what they write because you want them to answer what they honestly believe, not what they think you (their teacher) expects them to believe.

After collecting the students' papers, invite them to reflect with other students in the class on all of the activities in this unit. Did they change their minds about anything? If so, what was it that changed their minds? Was it something that somebody said? Or was it perhaps the opportunity to think about what the ball-shaped Earth concept really means?

You may want to conclude by pointing out the ways that the students have acted like astronomers in these activities:

CHAPTER 5: Earth's Shape and Gravity

- Explaining natural phenomena (day and night)
- Constructing mental models to explain phenomena they observe in the sky
- Performing thought experiments to test their models
- Using logic
- Referring to evidence
- Debating different alternative explanations
- Explaining why they believe what they believe

If your students have been doing these things, they are scientists doing science, and they can consider themselves to be astronomers.

Time for Reflection

For several years in the 1990s, this unit about Earth's shape and gravity was used as part of a summer institute for teachers from around the country. To evaluate the effectiveness of the program, we asked a group of about a dozen teachers to give pretests and posttests to students who participated in these classes. The results were gratifying. The study took place in 10 states with 18 classrooms and 539 students, ranging from grade 4 through grade 9. Although all groups significantly increased their scores on questions like those depicted in Figures 5.3–5.7, the youngest students, in grades 4 and 5, made the greatest gains. Students in control groups who did not engage in these activities did not increase their scores. In other words, students can advance in their understanding of the Earth's shape and gravity concept, even though it is a challenging topic to teach (Sneider and Ohadi 1998).

A sobering finding of our research was that not all of the students reached mastery, even at the highest grade levels. This finding highlights one of the tensions in today's educational landscape. On the one hand, we know that if our students are to learn to think creatively and critically and understand concepts at a deep level, they must be given opportunities to think through ideas and come to their own conclusions. Unraveling misconceptions takes time, and some students may only advance one or two levels of understanding, such as going from Notion 1 or 2, to Notion 3 or 4 as shown in Figure 5.1. On the other hand, some high-stakes tests measure mastery of concepts at just a few points in time, so they do not document incremental improvements in students' thinking. And when it comes to hard-to-teach concepts, such as the Earth's shape and gravity, our research shows that mastery is not easy to achieve.

What are the expectations in your school setting? Are your students expected to take high-stakes tests? If so, do the tests measure differences in levels of understanding as opposed to all-or-nothing gains in knowledge and skills? Are teachers rewarded for moving students along in their understanding from one year to the next? Or are teachers and students held to standards of excellence that are exceedingly difficult to reach?

The ideas in this book are intended to help you do a better job of teaching and make learning more enjoyable and successful for your students. And while they can-

Topic: The Force of Gravity (K–4)
Go to: *www.scilinks.org*
Code: HTT009

Topic: Gravity and Orbiting Objects (K–4)
Go to: *www.scilinks.org*
Code: HTT010

Topic: Gravity (5–8)
Go to: *www.scilinks.org*
Code: HTT011

Topic: Matter and Gravity (5–8)
Go to: *www.scilinks.org*
Code: HTT012

CHAPTER 5: Earth's Shape and Gravity

not help you resolve tensions between school policies and the realities of the classroom, a deeper understanding of what it takes to help students advance in their intellectual attainments may help you think through some of those dilemmas and celebrate your students' accomplishments. In my experience there is nothing more rewarding than to see a student's face when he or she realizes for the first time that people really do live "down there, on the other side of the world."

Responding to the Needs of All Learners

As discussed in Chapter 3, teachers can differentiate a lesson's content, process, and products. The following section summarizes some basic ways in which you can differentiate the content for these lessons.

Differentiating Content

Each student must understand the core content of the lessons. But the fifth lesson in this series of lessons addressed what is considered a "stretch" goal. It is considered a "stretch" goal because students are deemed proficient when they demonstrate solid understanding of the content of the first four lessons. I believe that all students can eventually complete this learning experience. However, if at this point (the fifth lesson) some of your students still struggle with the ideas in the lessons, you might consider this a point at which you differentiate instruction based on readiness.

There are several ways you can do this, but adjustable assignments (see Table 3.4, p. 142) are one good option. Although all students can benefit from adjustable assignments, consider this specific option for students who are still struggling. Your exact plans can vary depending on the number of proficient students, but let's assume you have only a couple of groups that are struggling. Adjust their assignments as follows:

1. Some students have a general understanding of the concepts but are struggling with vocabulary. The vocabulary is essential if students are to communicate their scientific understandings. Ask them to work in a group and develop operational definitions of the terms. Ask a proficient student to facilitate the work of this group.
2. Some students are still struggling conceptually. Group them with a proficient student to work on a concept map of the specific content that still challenges them. Proficient students can proceed with Lesson 5.
3. While proficient partners and groups are working together on Lesson 5, visit the groups set up in Step 2 and facilitate their work.

Differentiating Process

Chapter 3 outlined many ways in which you can differentiate aspects of a lesson. These included differentiation of learning activities, resources, grouping, learner needs, time, and space. Table 5.4 summarizes differentiation already found in the lessons in this chapter and possible further ways in which you can differentiate.

Table 5.4

Differentiation in the Instructional Plans for Learning Targets #1–4 (see Table 5.2, pp. 187–188)

	Present in the Instructional Plans for Learning Targets #1–4	**Possible Further Differentiation**
Learning Activities	• Common activities representing studied phenomena for this topic were selected but were modified by using research-based strategies (from the Instructional Tools, pp. 76–126) to better engage students in the learning process and probe their thinking. • Critical thinking is supported (e.g., students must provide a rationale for responses to questions and evidence for claims during inquiry). • Inquiry is infused into the lesson, giving students opportunities to draw on their learning strengths. • Various perspectives and cultures are represented in the different lessons.	Many strategies can be used to facilitate dialogue among groups. Consider using several different strategies during this series of lessons. One familiar example is Think-Pair-Share. Use the Chapter 5 figures (Figures 5.3–5.7) as the original prompt. Provide a minute or so for individual students to think about the prompt and determine their answers. Then give preassigned buddies (e.g., Clock Buddies) a few minutes to compare answers and come up with the response they think is best. Finally, call on random pairs to share their thinking. You can use craft sticks with students' names on them for random choice.
Resources	A variety of resources are used to learn targeted concepts: figures with images and text used to probe student thinking, paper and markers for drawing, three-dimensional models, online resources.	• Students can use Google Earth to fly around Earth, zoom in to visit a spot (possibly their school), and zoom back out and in to visit another place. At each place everything is "right-side up." Visit *http://earthi0-3d.com* • Use reading buddies or support from an instructional aide for struggling readers and English language learners. • Use audio versions of the prompts included in the figures. • Provide additional leveled reading materials for Lesson 6.
Grouping	Various groupings (partnered work, small groups, and whole class) are used at various times during the lesson. The small groups include students with varying degrees of understanding, as determined by student responses to each figure.	For some lessons, you might consider grouping by readiness to better facilitate the work of struggling students.

(continued)

CHAPTER 5: Earth's Shape and Gravity

Table 5.4 (continued)

	Present in the Instructional Plans for Learning Targets #1–4	Possible Further Differentiation
Learner Needs	• Student conceptions are elicited early in the lesson and thinking is made evident during the lesson through discourse, teacher questioning, and written responses. • A variety of intelligences are addressed through, for example, discourse, writing, drawing, concrete materials, or mock debates.	The lessons, while differentiated to some degree, still require students to work together in particular configurations (i.e., individual, partners, small group, or whole class) at particular times. Students work through the lessons at a pace predetermined by the teacher. Consider Lesson 5, which is a particularly flexible lesson.
Time	No adjustments are made in the plan.	Before introducing the unit, develop personal agendas for your students that are based on individual student learning needs. When some groups complete work earlier than others, ask them to work on a task in their personal agendas.
Space	Though various grouping configurations are used, no specific modifications for use of space are made in the plan.	Work with students to quickly shift furniture from whole-class to small-group configuration setups.

Differentiating Product

Lesson 6 suggests specific activities that serve as assessments because they require students to pull together what they have learned in previous lessons. The specific suggestions included (1) a writing assignment of a few paragraphs about a conversation with a member of the Flat Earth Society and (2) a mock debate. There are many other alternatives that address additional learning modalities. These could include a blog, a newsletter, expository writing, or a dramatic presentation to explain how students' thinking has changed. Your only limitation is that the assessments must align with the targeted content and established criteria to measure understanding.

Ties to Literacy and Numeracy (Mathematics)

The strongest ties to numeracy in this unit are in geometry (using visualization, spatial reasoning, and geometric modeling to solve problems), which is essential to modeling and a strong component of these lessons. Consider each of the following expectations from this National Council of Teachers of Mathematics geometry standard:

- Create and describe mental images of objects, patterns, and paths.
- Identify and build a three-dimensional object from a two-dimensional drawing.

- Identify and draw a two-dimensional representation of a three-dimensional object.
- Recognize geometric ideas and relationships and apply them to other disciplines and to problems that arise in the classroom or in everyday life.

If any of these standards are expectations at your grade level, consider how to connect them to the science lessons in this chapter.

Many strong ties exist between the science content in this chapter and literacy in writing and reading. The authors have used the draft Common Core Standards for the English Language Arts as their source for literacy standards; go to the Common Core homepage at *www.corestandards.org* and select the "English Language Arts Standards" link to find the English language arts standards and expectations that appear in Table 5.5.

Table 5.5

Science Connections to the English Language Arts

Language Arts Area	Specific Standards Area	Brief Description of Relevant Fifth-Grade Expectations	Science Connections in Chapter 5, Lessons 1–7
Reading Standards for Informational Text	Craft and Structure	Determine meanings of domain-specific words relevant to a fifth-grade subject area.	There is not an extensive vocabulary list for the lessons in Chapter 5, but certain words—*Earth, gravity, transparent, evidence, logic, phenomena*—are essential.
	Integration of Knowledge and Ideas	Use multiple sources to locate an answer to solve a problem; identify an author's use of evidence to support claims in text and identify the evidence that supports the claim; integrate information from various texts to write or speak about a subject.	These lessons rely on and develop students' abilities to give priority to evidence and logic in responding to questions, formulate explanations from evidence, and connect explanations to scientific knowledge.
Writing Standards	Text Types and Purposes	Write opinions and write informative/explanatory pieces.	Students write numerous explanations in these lessons. Lesson 6 requires students to do research about the Flat Earth Society and either write a dialogue between themselves and a society member or hold a mock debate.
	Research to Build and to Present Knowledge	Complete brief, focused research to build knowledge using various sources; gather relevant information from experience and print/digital sources and summarize information in written work; write in response to a focus question after investigating and discussing; write in response to informational sources, drawing evidence to support analysis.	

(continued)

CHAPTER 5: Earth's Shape and Gravity

Table 5.5 *(continued)*

Language Arts Area	Specific Standards Area	Brief Description of Relevant Fifth-Grade Expectations	Science Connections in Chapter 5, Lessons 1–7
Speaking and Listening Standards	Comprehension and Collaboration	Initiate and engage in discussions, coming prepared to draw from sources about the topic of discussion, contribute comments that build on the ideas of others, ask questions to clarify or follow up on ideas, and draw conclusions based on the discussion; summarize ideas and details presented orally, visually, or graphically; summarize claims made by a speaker and summarize the evidence used by the speaker to support the claims.	Discourse is essential in these lessons as students make sense of the science experiences. They are required to summarize and present their ideas throughout the lessons.
	Presentation of Knowledge and Ideas	Present an organized and logical report using appropriate and specific facts and details to develop main ideas (using visual displays as appropriate).	Lesson 6 depends on these abilities.

Consideration Across the Grades

These lessons are designed to work at either the fourth- or fifth-grade level. Teachers of younger students may use these lessons if they wish to stimulate class discussion, provided that they understand that most of their students are not likely to develop full understanding of the spherical Earth and gravity concepts.

Build Your Library

For Your Students

Books

Astronomy Activity Book (Dennis Schatz)
Flat Earth, Round Earth (Theresa Martin)
Gravity Is a Mystery (*Let's-Read-and-Find-Out Science* series) (Franklyn M. Branley)
How We Learned the Earth Is Round (Patricia Lauber)

For You

Astronomy Curricula

Astro Adventures (Dennis Schatz)
Astro Adventures II (Dennis Schatz)
Earth, Moon, and Stars (Cary I. Sneider)
The LHS GEMS Space Science Sequence, grades 3–5 and grades 6–8. (Lawrence Hall of Science, University of California, Berkeley)

Online Resource

Good Astronomy Activities on the World Wide Web (Andrew Fraknoi): *www.astrosociety.org/education/activities/astroacts.html*

Chapter 6

Understanding Changes in Motion

CHAPTER 6: Understanding Changes in Motion

> "Conservation of Motion in a Horizontal Plane... the subject provides contradictory explanations—light balls go farther because they are easier to set in motion. Larger ones go farther because they are stronger. There is an absence of laws."
>
> —*Inhelder and Piaget 1958*

Overview

During the late 1950s and early 1960s, several prominent physicists became deeply interested in elementary education reform. One of the most renowned reformers was Robert Karplus, a theoretical physicist from the University of California at Berkeley who was also very interested in the work of child psychologist Jean Piaget. One result of Karplus's work (and that of others) was the birth of the "Don't tell me, I'll find out" movement. This movement helped create elementary science "kits," a central part of the Elementary Science Study (ESS), the Science Curriculum and Improvement Study (SCIS), and the Science: A Process Approach (SAPA) programs. Later, in the early 1970s, there was a renewed call for elementary science education reform when it became apparent that many of the kits were no longer in use by teachers.

Around this time, two physicists, Arnold Arons and Lillian McDermott, came to the University of Washington and implored other scientists within the physics community to think more deeply about issues in science teaching. They closely examined science teaching at the elementary school and realized that teaching for understanding requires considerable teacher expertise. They also found that certain conceptual difficulties were not easily addressed, even by the best curriculum. As a result they called for a shift in focus from producing materials for children to creating more effective materials and courses for teachers—particularly elementary school teachers. To help identify and address these issues, McDermott created the Physics Education Group and began to conduct research to uncover specific difficulties that students encounter while learning foundational ideas in physical science. The result was the research-based materials for teachers called Physics by Inquiry (McDermott and the Physics Education Group at the University of Washington 1996).

One of the first topics examined by researchers in McDermott's Physics Education Group was motion. They conducted extensive interviews with students and teachers and discovered a vast array of conceptual stumbling blocks, most of which centered around two areas: (1) representations of motion (e.g., interpreting motion graphs) and (2) differentiating between a quantity and a change in that quantity (McDermott and Rosenquist 1987). They discovered that these stumbling blocks were far more common than they had expected among both children and adults and that a traditional approach to teaching these topics was strikingly ineffective.

CHAPTER 6: Understanding Changes in Motion

To illustrate just how common these difficulties are, consider the following transcript from a speeding trial (Lochhead 2010). A driver is arguing to the judge that the officer had mistaken his car for another car that was passing him.

> Judge: If the two cars were side by side, you were both speeding.
> Defendant: The car was passing me; I was not speeding.
> Judge: I am satisfied that the weight lies with the commonwealth in this case…

In this example, the judge had confused position (location) with speed. Researchers at the University of Washington designed an interview protocol around this difficult topic called the Speed Comparison Task (Trowbridge and McDermott 1981a). Their study showed that roughly 60% of adults with no background in physics had difficulty understanding the difference between location and speed and between 10% and 30% of physics students, even after instruction, also experienced this difficulty. Trowbridge and McDermott (1981b) also conducted research to identify difficulties that students had in understanding the difference between velocity and acceleration (acceleration is the time rate of change of velocity). These results were even more striking. After instruction, 80% to 90% of adults with no background in physics and over 60% of physics students showed significant difficulties differentiating between these two concepts. Based on this evidence, it is not surprising that students struggle to understand Newton's second law, which relates the net force acting on an object to the acceleration of that object. As a result, acceleration (a change in velocity in a given time interval) should be considered an important prerequisite for understanding Newton's laws of motion. In the draft conceptual framework for the Next Generation Science Education Standards (NRC 2010), these ideas are reflected in "learning progressions" and are sometimes referred to as "stepping stones."

This chapter describes how to apply a conceptual change framework to help children understand motion and changes in motion. These lessons can be taught at any of the third- through fifth-grade levels, but the lessons as described target third grade. Fourth- and fifth-grade modifications are shared in the Consideration Across the Grades section of this chapter, page 230. The essential question is "How do we observe and then describe *change*?" Specifically, we are interested in what type of learning should occur in the early grades to prepare students for the more sophisticated notions of force they will learn as they get older.

Why This Topic?

Force and motion are difficult topics for many reasons, including students' preconceptions regarding the relationship between force and motion and a young student's limited access to symbolic tools as they relate to such ideas as ratios, rates, and fractions. (See Table 1.1, p. 5, for the five major reasons that certain science concepts can be

CHAPTER 6: Understanding Changes in Motion

difficult to learn and teach.) In addition, these difficulties are related to a broader class of difficulties involving proportional reasoning and have been identified as problematic in many different contexts, including the fields of biology (growth rates), geography (scaling), and economics (price or cost of goods and interest rates). As a result, students who do not have an opportunity to develop ideas about force and motion at a young age will continue to experience difficulties as they get older.

Some teachers argue that the introduction of topics related to ratios and proportions be delayed until students are older and more developmentally ready. However, researchers have found that *more* exposure to challenging ideas, not *less*, is required for students to begin to develop good thinking skills. Researchers Phillip Adey and Michael Shayer conducted an extensive study spanning 30 years in which they studied the effects of introducing cognitively challenging activities to children as young as four years old (Shayer and Adey 2002). They found striking results: Students who were exposed to cognitively challenging experiences starting in the primary grades showed a measurable gain in ability throughout their school careers. Consistent with this result is the National Research Council (NRC) report titled *Taking Science to School: Learning and Teaching Science in Grades K–8* (Duschl, Schweingruber, and Shouse 2009). The report states that the commonly held view that young children are concrete and simplistic thinkers is outmoded; research shows that children's thinking is surprisingly sophisticated. Yet much current science education is based on the old assumptions and so focuses on what children cannot do rather than what they can do. Children can use a wide range of reasoning processes that form the underpinnings of scientific thinking, even though their experiences are variable and they have much more to learn.

> Contrary to conceptions of development held 30 or 40 years ago, young children can think both concretely and abstractly.... Development is not a kind of inevitable unfolding in which one simply waits until a child is cognitively "ready" for abstract or theory-based forms of content. Instead, parents and teachers can assist children's learning, building on their early capacities. (Duschl, Schweingruber, and Shouse 2009, p. 3)

In this chapter, we describe one approach to introducing these difficult ideas to third-grade students using the conceptual change framework (the Instructional Planning Framework, Figure 1.2, p. 8).

What Makes These Ideas Difficult?

Many of the challenging concepts that students encounter in science are related to difficulties in understanding proportion and ratios. Changes in a quantity with respect to time are called *rates*. There are many examples of quantities and their rates: debt and deficit, population and growth rate, position and speed, and electric charge and electric current. Differentiating between quantities and their time rate of change is critical to understanding many important ideas in a wide range of disciplines.

CHAPTER 6: Understanding Changes in Motion

Considerable research supports the idea that ratio reasoning is difficult for students at all grade levels (Arons 1983, 1984). In addition to the conceptual difficulties that students encounter with ratios, students face deep conceptual challenges related to understanding the relationship between force and motion. Isaac Newton, who first challenged the idea that motion requires a force, proposed the idea that only a *change in motion* requires a force. Research provides strong evidence of the difficulties associated with these ideas. Even college students (close to 50%) do not change their views on these ideas even after taking a traditional college-level physics course (Hake 1998).

These difficulties are the result of years of experience with motion in our daily lives. These experiences reinforce the everyday belief that a constant motion results from applying a constant force. For example, a book will slide across the table only if it is being pushed, or you can run across the playground only if you move your legs. It is a challenge to teach students at any age that there are hidden forces that cannot be observed directly (in these examples, the friction between the book and the table or between your feet and the playground) and that in the absence of these hidden forces, motion at constant speed would require no effort at all (Champagne, Klopfer, and Anderson 1980). Even with concerted efforts, it can still be difficult (if not impossible) to convince students that if friction forces were taken away, a moving book would continue to move without slowing down or that you could slide at a constant speed across the playground without moving your legs. To begin to understand the relationship between force and motion, students must understand what it means for motion to change and be able to describe changes in motion.

The Predictive Phase

In the first stage of the framework, our task is to identify the important ideas that should be the target for each lesson. We have found that Curriculum Topic Study (CTS) Guides can be useful resources for this phase of planning (Keeley 2005). These guides help teachers identify key topics and connect these topics to related research on student difficulties and to related National Science Standards (NRC 1996) and the Benchmarks for Science Literacy (AAAS 1993). The CTS guides related to force and motion include "Motion," "Describing Position and Motion," "Laws of Motion," and "Forces." For the purpose of this chapter, we will focus on ideas related to motion and descriptions of motion. To help connect the lessons with the target ideas related to understanding changes in motion, we have used the planning template introduced in Chapter 2 (Table 2.6, p. 34).

The planning template in this case helps to connect specific instructional strategies to learning targets and standards. In many cases, the learning targets and their specific instructional plans may, and usually do, overlap. However, identifying the essential goals and enduring understandings of a unit prior to working on specific lesson plans or classroom teaching is an important prerequisite for effective instruction (see, for example, *Understanding by Design* by Grant Wiggins and Jay McTighe 1998). The completed planning template for the predictive phase for lessons on motion is shown in Table 6.1 (p. 210).

CHAPTER 6: Understanding Changes in Motion

Table 6.1

Predictive Phase: Planning Template for "Understanding Changes in Motion"

	Lesson Topic: "Understanding Changes in Motion"
Essential Understandings	Objects can move fast, move slow, or be at rest. Objects can also change their speeds. When an object speeds up or slows down, then the object is changing its speed. Changing speed is evidence that the object is interacting with other objects.
Knowledge Required From Previous Instruction	• Position is a description of where an object is located. • Distance is how far an object can travel (also referred to as *displacement* in later grades). • Length can be measured with a ruler or a meterstick. • A clock can be used to tell us what time it is. • A stopwatch can be used to measure how long (the amount of time) it takes for an event to occur.
Knowledge and Skills to Be Learned	**Concepts**: See the Learning Targets below. **Vocabulary:** *distance, time, speed, fast, slow, faster, slower, speeding up, slowing down* **Skills:** Measuring length (ruler or meterstick), measuring time intervals (stopwatch).
Criteria to Demonstrate Understanding	• Be able to compare the speeds of two objects (which object is moving faster than the other object). • Be able to provide evidence that the speed of an object is not changing. • Be able to provide evidence that the speed of an object is changing. • Be able to tell the difference between "how long it takes," "how far it has traveled," and "how fast it is moving."
Learning Targets and Instructional Plans	
Target #1	*Objects can change speeds. How long it takes for an object to move between two points is not the same as asking how fast an object is moving.*
	Instructional Plan for Learning Target #1: Will be completed during the responsive phase (Table 6.3)

Table 6.1 (continued)

Target #2	Object A is moving faster than Object B if Object A travels the same distance in a smaller amount of time or if Object A travels a longer distance in the same amount of time.
	Instructional Plan for Learning Target #2: Will be completed during the responsive phase (Table 6.3, p. 218).
Target #3	If an object changes the distance it travels in equal amounts of time, then the object is either speeding up or slowing down.
	Instructional Plan for Learning Target #3: Will be completed during the responsive phase (Table 6.3).
Target #4	Measurements of time and distance can provide evidence for knowing if an object is speeding up, slowing down, or moving at the same speed.
	Instructional Plan for Target #4: Will be completed during the responsive phase (Table 6.3)

Stage I: Identify the Conceptual Target

(For a complete list of the ten stages for implementing the Instructional Planning Framework, see Table 2.1, p. 19.)

Stage II: Unpack the Standards and Identify the Concepts, Knowledge, Skills, and Vocabulary for the Content You Are Teaching

For this example, we used the results of research on student difficulties as the starting point for our conceptual targets. To connect the results of research with specific instructional goals, we then closely examined the relevant parts of the National Science Education Standards and the Benchmarks for Science Literacy. It is useful to examine those standards that relate to students who are younger and older than those we teach. (*Note:* Resources that are particularly useful for this work are the two volumes of the *Atlas of Science Literacy* [AAAS 2001b, 2007].)

The section of the National Science Education Standards (NRC 1996) that relates directly to understanding motion is located in Physical Science Content Standard B: Position and Motion of Objects. The K–4 standards include the following:

- The position of an object can be described by locating it relative to another object or the background.
- An object's motion can be described by tracing and measuring its position over time.
- The position and motion of objects can be changed by pushing or pulling. The size of the change is related to the strength of the push or pull. (p. 127)

CHAPTER 6: Understanding Changes in Motion

These standards lead into the grades 5–8 standard:

- The motion of an object can be described by its position, direction of motion, and speed. (p. 154)

In *Benchmarks for Science Literacy* (AAAS 1993), the chapter "The Physical Setting" has a subsection specifically titled "Motion," in which the following Benchmarks appear:

By the end of the 2nd grade, students should know that

- Things move in many different ways, such as straight, zigzag, round and round, back and forth, and fast and slow.
- The way to change how something is moving is to give it a push or a pull.

By the end of the 5th grade, students should know that

- Changes in speed or direction of motion are caused by forces. The greater the force is, the greater the change in motion will be. The more massive an object is, the less effect a given force will have.
- How fast things move differs greatly. Some things are so slow that their journey takes a long time; others move too fast for people to even see them.

By the end of the 8th grade, students should know that

- An unbalanced force acting on an object changes its speed or direction of motion, or both. (pp. 89–90)

Benchmarks for Science Literacy also contains a chapter called "Common Themes." One of the common themes (referred to as "Cross-Cutting Elements" in the new Common Core State Standards Initiatives [2010]) is "Constancy and Change." The following Benchmarks are from that section:

By the end of the 2nd grade, students should know that

- Objects change in some ways and stay the same in some ways.
- People can keep track of some things, seeing where they come from and where they go.
- Small changes can sometimes be detected by comparing counts or measurements [at different times].
- Some changes are so slow or so fast that they are hard to [notice while they are taking place]. (p. 272)

By the end of the 5th grade, students should know that

- Some features of things may stay the same even when other features change.

- Things change in steady, repetitive, or irregular ways—or sometimes in more than one way at the same time. Often the best way to tell which kinds of change are happening is to make a table or graph of measurements. (p. 173)

The ability to differentiate between a quantity and a change in a quantity is implicit in the Standards and Benchmarks at every grade level. The beginning of this understanding rests on observation and language. As students progress, they learn to quantify these changes through measurement and mathematical representations.

Stage III: Identify and Sequence the Subgoals (the Learning Targets)

We have identified four learning targets that build on one another. These learning targets are based on research into student thinking related to the concepts of motion (see Stage VI). Additional details about each learning target are contained in the sections that follow.

- Learning Target #1: Objects can change speeds. How long it takes for an object to move between two points is not the same as asking how fast an object is moving.
- Learning Target #2: Object A is moving faster than Object B if Object A travels the same distance in a smaller amount of time or if Object A travels a longer distance in the same amount of time.
- Learning Target #3: If an object changes the distance it travels in equal amounts of time, then the object is either speeding up or slowing down
- Learning Target #4: Measurements of time and distance can provide evidence for knowing if an object is speeding up, slowing down, or moving at the same speed.

Stage IV: Identify Criteria for Determining Student Understanding

In all four of the lessons we describe later in this chapter, students use evidence-based reasoning. We have found that students often use words that might indicate understanding, but these words can too often be used to disguise a lack of conceptual understanding. As a result, the primary criteria for assessing student performance should be *functional understanding*, defined as an ability to apply a concept learned in one context to a new context. This requires students to show their operational knowledge (as compared to declarative knowledge) and requires students to show the teacher and one another that they understand the difference between motion and changing motions (speeding up or slowing down). Acting out "the story" can also be illuminating. For example, a prekindergarten class of four-year-olds had experimented with marbles rolling along different paths they had constructed. After this activity, the teacher asked students to show him what it means to "be a marble." The teacher then had students draw the path they had taken during a walk in the woods in the same way he had asked students to draw the path of the marble during their experiments (Gilsdorf and Loftus 2010).

As students get older, they can be given specific tasks such as "Here is a ball rolling across the table. Show me how you would find the speed of the ball using a stopwatch and a meterstick." Such tasks yield different information about student thinking than would be obtained by asking students directly for various definitions.

Stage V: Identify Inquiry and Metacognitive Goals and Strategies

It is critically important that the teacher creates a safe classroom environment that encourages students to share their ideas. Asking students to be metacognitive means asking them to think about their own thinking. This is a difficult proposition that begins with asking students to describe why they believe what they believe. *Nonlinguistic* metacognitive strategies include having the students draw pictures (such as a cartoon) or "act out" their thinking. Drawing "change" can be difficult, and asking students to "draw the motion" can yield a variety of representations, such as "snapshots" (e.g., a series of pictures to show how change occurs) or lines or arrows showing "motion" at an instant.

Sometimes, a talk-out-loud strategy, rather than a nonlinguistic strategy, can be more effective. For example, students might perform a task while they narrate what they are doing and why they are doing it (see Instructional Tool 2.10, p. 125). If a student were to, for example, use the words "going faster," the teacher could ask the student to show what he or she means by running in a way that demonstrates "going faster." The teacher could then ask another student to show "speeding up" to see if students are thinking that these words mean the same thing or something different.

Motion experiments work well in many forms of scientific inquiry. In the four lessons described later in the chapter, teachers can use two of the five essential features of inquiry discussed in Instructional Tool 2.1, p. 76): Formulate Explanations From Evidence and Give Priority to Evidence in Responding to Questions. For example, if a student says, "The ball is speeding up," one response from the teacher could be, "Tell me how you know it is speeding up. What is it that you are seeing or hearing?" Another good response to student questions would be to suggest a class experiment that would provide evidence to support or refute an idea. Direct answers to questions tend to hinder the inquiry process because they encourage students to see the teacher as the source of all information rather than to use experimental results and their own direct observations.

The Responsive Phase

In this section, we discuss several lessons on motion that can be used with third-grade children. The important idea is to base these lessons directly on what was learned during the predictive phase and to pay close attention to the various strategies that are required for conceptual change.

Stage VI: Research Children's Misconceptions Common to This Topic That Are Documented in the Research Literature

Understanding children's ideas related to motion has been the subject of researchers for many years. For an overview of this research, see Driver et al. (1994) and McDermott and Rosenquist (1987). In addition to the difficulties described in the Overview section, researchers have found that children tend to see objects in two categories: at rest or moving. This way of seeing is sometimes referred to as a "snapshot view," in which students do not refer to time in their descriptions (Dykstra and Sweet 2009). For example, in a snapshot view, students do not readily differentiate between the words *fast* and *going faster*. Describing an object as "going faster" is quite often ambiguous, and teachers need to distinguish between when a student is comparing the speeds of two objects (one is going "faster" than the other) and when he or she is describing how the speed of an object is changing ("speeding up" or "slowing down"). Children often use the phrase "going faster" simply to mean that the object is "fast." Another use of the word *faster* relates directly to a child's concept of time. In this context, *faster* means "less time" (e.g., this playground slide is "faster" than that playground slide).

One way to organize student ideas is to identify "facets" or possible ways that students think about motion (Minstrell et al. 2008). As described by Minstrell (2000), "Facet based instruction centers around the idea that students, faced with a problem situation, apply preformed ideas from previous experiences or construct ideas and reasoning to make sense of the situation." Minstrell and his colleagues have identified over 130 different facets related to motion and changes in motion. Many of these facets are related to calculation errors that may or may not be associated with an underlying conceptual difficulty. However, the primary difficulty at the root of many student ideas is related to a confusion regarding how a quantity is related to a change in that quantity. (For a list of facet codes, see *http://depts.washington.edu/huntlab/diagnoser/facetcode.html*.)

Stage VII: Select Strategies to Identify Your Students' Preconceptions

Stage VIII: Select Strategies to Elicit and Confront Your Students' Preconceptions

Stage IX: Select Sense-Making Strategies

Conceptual change requires specific strategies to help students change the way they think about a particular situation. For the lessons on motion, we have used a variety of approaches but will focus on the instructional strategy called Kinesthetic Activities (Instructional Tool 2.10, p. 125, and Instructional Tool 2.9, p. 121). In addition, we will

CHAPTER 6: Understanding Changes in Motion

Table 6.2

Strategy Selection Template for "Understanding Changes in Motion"

	Learning Target #1	**Learning Target #2**	**Learning Target #3**	**Learning Target #4**
	Objects can change speeds. How long it takes for an object to move between two points is not the same as asking how fast an object is moving.	Object A is moving faster than Object B if Object A travels the same distance in a smaller amount of time or if Object A travels a longer distance in the same amount of time.	If an object changes the distance it travels in equal amounts of time, then the object is either speeding up or slowing down.	Measurements of time and distance can provide evidence for knowing if an object is speeding up, slowing down, or moving at the same speed.
Possible Strategies for:				
Identifying Preconceptions	Specific strategies to identify student preconceptions were not used. Instead, the development of the unit was based on student misconceptions that had already been identified in the research.			
Eliciting and Confronting Preconceptions	• Small- and Large-Group Discourse • Hands-on Experiments and Activities and Manipulatives	• Physical Movement • Small- and Large-Group Discourse • Hands-on Experiments and Activities and Manipulatives	• Physical Movement • Small- and Large-Group Discourse • Hands-on Experiments and Activities and Manipulatives	• Small- and Large-Group Discourse • Small-Group Discourse (Instructional Tool 2.6, Speaking to Learn) • Hands-on Experiments and Activities and Manipulatives
Sense Making	Hands-on Experiments and Activities and Manipulatives	Hands-on Experiments and Activities and Manipulatives	Drawings and Annotated Drawings	Hands-on Experiments and Activities and Manipulatives
Demonstrating Understanding	Student Questioning	Student Questioning	Small-Group Discourse (Instructional Tool 2.6, Speaking to Learn)	Hands-on Experiments and Activities and Manipulatives

Table 6.2 (continued)

Selected Inquiry Strategies	1. Formulate Explanations From Evidence 2. Give Priority to Evidence in Responding to Questions Explanation includes *making a claim, supporting the claim with appropriate and sufficient evidence,* and *reasoning,* which links the claim with evidence and explains why the data are evidence in support of the claim (McNeill and Krajcik 2008).
Selected Metacognitive Strategies	1. Drawing Out Thinking 2. Self-Regulate Students externalize their thinking, thinking out loud, using a variety of strategies. By verbalizing their thinking, students gain awareness and control over their problem solving and gain a fresh perspective on their own thoughts.

discuss the metacognitive strategy Self-Regulate as described in Instructional Tool 2.2, p. 83. See Table 6.2, p. 216, for a summary of the strategies used in the lessons on motion.

As one teacher told us, the study of motion is perfect for young children because "They are always in motion themselves!" When teaching about motion, a great deal can be accomplished under the guise of "play," with students running, jumping, spinning, and dancing. Providing students with a few marbles or tennis balls along with materials to construct various paths for the balls, such as ramps or tracks, can be a wonderful way to introduce motion before starting to ask the "what happens if" questions required to extend their thinking. Asking questions naturally leads to conducting experiments and is an opportunity to encourage students to use evidence-based reasoning.

After the direct kinesthetic experiences, students should be encouraged to represent their ideas with words and on paper. At this age, students should draw pictures to show what they observed, and teachers should identify the different ways that a child attempts to draw "motion." This strategy, Drawing Out Thinking, can also be deeply metacognitive as students must reflect on what they observed to construct their drawings.

The specific metacognitive strategy that is used in most of the lessons on motion is called Self-Regulate. This form of self-evaluation arises from having a self-identified criteria (making a prediction), performing an experiment, and then using the results of the experiment to make a judgment about the original prediction. At this stage, students are encouraged to compare their thinking and to discuss and record the differences between what they thought would happen and what actually happened.

Stage X: Determine Formative Assessments

Many of the instructional strategies in the four lessons in this chapter involve hands-on experiments and therefore lend themselves to formative assessments based on direct observations. Student thinking can be further clarified through teacher questions during each of the activities. Examples of specific questions (called *probes*) designed to

CHAPTER 6: Understanding Changes in Motion

elicit student ideas related to motion can be found in the book *Uncovering Student Ideas in Physical Science, Vol. 1: 45 New Force and Motion Assessment Probes* (Keeley and Harrington 2010). Some of the lessons in this chapter are based on these probes. To ensure accurate assessments, students should record their own ideas before sharing their ideas with others. The teacher might hand out 3 × 5 note cards and ask students to write or draw their ideas prior to a classroom discussion. Collecting these cards immediately (before the class discussion) and reading quickly through them can help the teacher guide the discussion because he or she now has a good understanding of what students are thinking before the activity begins.

Once we identified the topics of the four lessons and selected appropriate strategies, we planned the lessons. The Responsive Phase Planning Template (Table 6.3) outlines the lessons.

Table 6.3

Responsive Phase: Planning Template for "Understanding Changes in Motion"

Identifying Student Preconceptions	• Refer to research on student difficulties with motion (Driver et al. 1994; Dykstra and Sweet 2009; McDermott 1996). • Each lesson begins with an elicitation of student ideas (use the Drawing Out Thinking strategy).

Learning Target #1:
Objects can change speeds. How long it takes for an object to move between two points is not the same as asking how fast an object is moving.

Instructional Plan for Learning Target #1
(This plan does not account for changes that will come about during instruction or in re-teaching.)

Elicit Preconceptions: Collect ideas and predictions for "Rolling Marbles" probe.
Confront Preconceptions: Perform experiment, students observe the result.
Sense-Making Experiences: Group discourse and then student groups construct their own experiments.
Using Formative Assessments: Use "Rolling Marbles" probe from *Uncovering Student Ideas in Physical Science, Vol.1: 45 New Force and Motion Assessment Probes* (Keeley and Harrington 2010).

Learning Target #2:
Object A is moving faster than Object B if Object A travels the same distance in a smaller amount of time or if Object A travels a longer distance in the same amount of time.

Table 6.3 *(continued)*

Instructional Plan for Learning Target #2
(This plan does not account for changes that will come about during instruction or in re-teaching.)

Elicit Preconceptions: Students share their ideas during the motion safari walk.

Sense-Making Experiences and Using Formative Assessments: Teacher directs students to observe specific motions and then ask students to decribe those motions (speeding up, slowing down, or moving at constant speed).

Learning Target #3:
If an object changes the distance it travels in equal amounts of time, then the object is either speeding up or slowing down.

Instructional Plan for Learning Target #3
(This plan does not account for changes that will come about during instruction or in re-teaching.)

Elicit Preconceptions: Use the Speed Comparison Task and ask students to identify when the two balls have the same speed.

Confront Preconceptions: Use two balls, each rolling at constant but different speeds, to show that when the balls pass, they do not have the same speed.

Sense-Making Experiences: Have students act out the Speed Comparison Task.

Using Formative Assessments: Group and individual discussion during the Speed Comparison Task provides formative information regarding student ideas. Repeat the Speed Comparison Task and have students record their predictions on note cards.

Learning Target #4:
Measurements of time and distance can provide evidence for knowing if an object is speeding up, slowing down, or moving at the same speed.

Instructional Plan for Learning Target #4
(This plan does not account for changes that will come about during instruction or in re-teaching.)

Elicit Preconceptions: Posing the question "Can you create uniform motion?" will elicit pre-existing ideas from students prior to the experiment.

Confront Preconceptions: Experimental results will provide evidence that answers the essential question of the lesson.

Sense-Making Experiences: Group discussion on the results of the experiment.

Using Formative Assessments: Use "Just Rolling Along" probe from *Uncovering Student Ideas in Physical Science, Vol.1: 45 New Force and Motion Assessment Probes* (Keeley and Harrington 2010).

CHAPTER 6: Understanding Changes in Motion

The Lessons: Teaching and Learning About "Understanding Changes in Motion"

Lesson 1: Eliciting Student Ideas About Speed, Changing Speed, and Time

The first lesson involves predicting which of three marbles will reach the bottom of a ramp first. There are three ramps and each ramp has a different shape (see Figure 6.1). All three ramps start at the same height. The primary goal of the exercise is to note the different words that students use to describe motion and to gain insight into *how they use* these words.

Figure 6.1

Three Different Ramps

Below are comments from students in a classroom during the elicitation phase of the lesson.

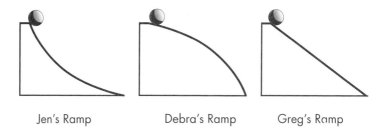

Jen's Ramp   Debra's Ramp   Greg's Ramp

- "I think that Jen's ramp will go the fastest because it starts by going downward, which causes the marble to roll faster because of the force and other stuff."
- "I think they [Debra's and Jen's marbles] have to go over a hump. Greg's is directly down so it will have more speed."
- "I think Jen's ramp will be the fastest because it is more downhill than the other ramps."
- "Because Debra's ramp and Jen's ramp are the same ramp but flipped. So Debra's will go slow then fast. And Jen's will go fast then slow. And Greg's will go at a steady pace just enough to make it a three-way tie."
- "Well I thought of skateboarding and in skateboarding Debra's ramp would be the fastest."
- "Because the fastest way to get somewhere is to go in a straight line, and Greg's ramp is straight."

CHAPTER 6: Understanding Changes in Motion

Following the elicitation phase, students should conduct tests with the ramps. In some cases, it is appropriate to discuss ideas regarding "what makes a test fair" and to gain a classroom consensus about this idea before testing. In other cases, it might be more appropriate for the teacher to perform the tests while students observe and share their ideas with one another. Later, even the youngest students will enjoy constructing different-shaped ramps and holding races to see which marble will take the longest and which will take the shortest amount of time to reach the bottom.

Although some students may notice differences in the shapes of the ramps, the primary goal for young children is to gain exposure to the language of motion (*slow, fast, speeding up, slowing down*) and to connect the language to real motions. In addition, students are also exposed to the idea of constructing and conducting a test. Teachers should be alert to students who use the word *fastest* to denote which ramp results in the marble reaching the bottom first. In the ideal case, if each marble begins at the same height, then each marble will be moving at the same speed when it reaches the bottom, even though some marbles will take longer to reach the bottom than others (Keeley and Harrington 2010).

Lesson 2: Observing Motions in Everyday Life

In Lesson 2, students are asked to observe motions that occur all around them. In a lesson that elementary school teacher Dennis Gilsdorf has called a "motion safari," teachers take students on a walk around the school to observe all the different types of motions that they can see. During the walk, students can practice using words such as *fast, slow, speeding up,* or *slowing down.*

Back in the classroom, students share their findings with the class and act out the motions that they observed. "Acting out" difficult concepts can be a powerful method for making student ideas more visible (Wilhelm and Edmiston 1998). These mini-dramas can be used as a point of discussion to help students understand the difference between going fast and speeding up.

Lesson 3: The Speed Comparison Task

In this lesson, the primary activity is based on what has been called the Speed Comparison Task. This task was the focus of a research study conducted at the University of Washington by David Trowbridge and Lillian McDermott (1981). It involves rolling two marbles side by side. One marble is moving at a constant speed and the other is speeding up. Students identify when they believe the two marbles have the same speed. The task is constructed so that the marbles pass each other twice. The marble that is rolling at a constant speed starts behind the marble that is released from rest on an incline (see Figure 6.2, p. 222). For specific instructions on how to construct these ramps see *Physics by Inquiry, Volume 1* by McDermott and the Physics Education Group at the University of Washington (1996).

CHAPTER 6: Understanding Changes in Motion

Figure 6.2
Speed Comparison Task

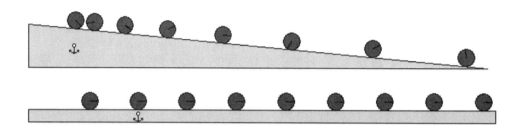

Many students believe that the marbles move at the same speed when they pass one another. This misconception is the result of not differentiating between position and speed. It is useful for third-grade students to see both motions side by side as this provides an opportunity to discuss any differences they might see. This lesson addresses Learning Target #1 ("Objects can change speeds. How long it takes for an object to move between two points is not the same as asking how fast an object is moving.")

Following this exercise, a second motion safari can be conducted during which students look around their school or homes to identify motions that are changing.

Lesson 4: Making Uniform Motion

The final lesson is a way to extend student thinking beyond observation and qualitative descriptions. In this lesson, students are encouraged to think about how they could use a ruler or a stopwatch to measure a motion. These skills should build on other lessons students receive on these topics, most notably in their math curriculum (time, clock readings, measuring length). The central task of this lesson is for students to create uniform motion with a marble and a table or metal U-shaped track that can be tilted. The teacher asks, "Is it possible to create uniform motion? Can you roll a marble so that the marble does not slow down or speed up?" A preassessment can be conducted using the "Just Rolling Along" probe from *Uncovering Student Ideas in Physical Science, Vol. 1: 45 New Force and Motion Assessment Probes* (Keeley and Harrington 2010).

Drama can also be brought into this lesson by calling each group a science research team and giving the groups names (such as the name of a university or a country). Students can also be referred to as "Dr. Susie" or "Professor David" and given the task of conducting an experiment and then presenting their results at a scientific conference, an exercise that helps teach students to "think like a scientist."

Student presentations to the class at the end of the lesson should include evidence for students' reasoning. Although some third graders may use measuring instruments such as stopwatches and rulers, many will rely solely on what they see or the sounds that they hear (e.g., students often notice that the sound of a rolling ball will change in pitch as the ball speeds up or slows down). Regardless, presentations can be a useful method of assessing student understanding of motion and can help a teacher identify those students who have started to correctly differentiate between constant motion and changing motion.

Time for Reflection

There are many examples of quantities and changes in quantities both in and out of our classrooms. As students work through each of the motion lessons, they should make connections with other parts of the curriculum and with their lives outside of school. What other changes are important in our lives? Can the rate of change also change? For example, am I growing "faster" than my older sibling? Do we have the same change in the amount of daylight today as yesterday? Or is the amount of daylight we experience each day changing? Can we make analogies between changes in daylight and changes in speed? How can we record and then represent these changes? Because all of us are more comfortable learning new ideas in contexts we are most familiar with, a good question to ask yourself is, "In what context do my students understand change?"

As students come to understand what it means for a motion to change, they also become ready to connect this observation to the idea of force (interactions with other objects). For example, the observation that an object speeds up when dropped from rest is evidence of the interaction with the Earth that we call *gravity*. The observation that a book slows down is evidence that the table is interacting with the book due to a force called *friction*. Eventually students learn that all changes in motion are the result of interactions with other objects.

Are there specific ways in which you can focus instruction to help students understand change? Consider what these ways might be and make as many connections as you can during the school year.

Responding to the Needs of All Learners

We now consider how to differentiate the content, process, and products of the lessons in this chapter.

Differentiating Content

So what happens if students do not understand the targeted concepts at the end of one of the lessons? You know that each student should understand the key concepts in each of the learning targets and that you do not want to expect less of them in terms of core content. Review Table 6.4, page 224, for some initial ideas of how you might address the content learning needs of struggling students while challenging those students who have already demonstrated understanding of key ideas.

CHAPTER 6: Understanding Changes in Motion

Table 6.4

Differentiation in the Instructional Plans for "Understanding Changes in Motion"

Lesson	Possible Differentiation During Initial Instruction	Options for Re-teaching
#1–Eliciting Student Ideas About Speed, Changing Speed, and Time	Because vocabulary development is essential in this lesson, you might provide struggling students with some instruction on the concepts prior to reading. For English-language learners, provide visual images and concrete examples of the terms. You can also extend vocabulary for students who demonstrate readiness.	For students who do not relate the vocabulary to the actual phenomena, you can provide additional experiences with the phenomena and use vocabulary development strategies that are effective with various learning styles. Allow students who have mastered the use of the vocabulary words to conduct additional experiments.
#2–Observing Motions in Everyday Life	Assign roles to students based on their learning styles as they prepare the dramatic presentation. Those less comfortable with acting can write scripts or sketch drawings that outline the steps of the production. Others can check the production to make certain it aligns with the group's best explanation for the concept.	Partner students who have mastered the content with those who are still struggling. Have them coach their partners on the various concepts and work with them to develop representations of the concepts in ways that differ from the dramatic production. Perhaps they can gather a photo collection that represents the concept and display the collection along with a written description of why the photos represent the concept.
#3–The Speed Comparison Task	See specific suggestions for differentiating process in Table 6.5.	Work with small groups of students who have not yet mastered the concepts. Have advanced students generate questions they still have and conduct investigations to test those questions.
#4–Making Uniform Motion	Differentiation is already embedded since some students present quantitative evidence to support results while others rely on what they see or hear.	

Differentiating Process

Let's consider ways in which we might differentiate learning activities, resources, grouping, learner needs, time, and space. Table 6.5 summarizes differentiation already found in the lessons and further ways in which you can differentiate.

Table 6.5

Summary of Differentiation for "Understanding Changes in Motion"

	Present in the Instructional Plans for Learning Targets #1–4	**Possible Further Differentiation**
Learning Activities	• Common activities representing studied phenomena for this topic were selected but modified by using research-based strategies (from the Instructional Tools) to better engage students in the learning process and probe their thinking. • Critical thinking is supported (e.g. students must provide a rationale for responses to questions and evidence for claims during inquiry). • Inquiry is infused in the lesson, giving students opportunities to draw on their learning strengths and providing choice in experimental design. • In Lesson #1, the author suggests options for the tests—either demonstrated by the teacher or designed by the students.	See suggestions in Table 6.4.
Resources	A variety of resources are used to learn targeted concepts: equipment for the various tasks, neighborhood resources for "motion safari," and measuring devices.	If teaching aides are available, have them facilitate some of the groups.
Grouping	No clear grouping practices are described.	Consider heterogeneous grouping during initial instruction and grouping based on readiness if re-teaching is required.

(continued)

Table 6.5 *(continued)*

	Present in the Instructional Plans for Learning Targets #1–4	**Possible Further Differentiation**
Learner Needs	• Student conceptions are elicited early in the lessons and thinking is made evident throughout the lesson, primarily through discussion. • A variety of intelligences are addressed through use of discourse, hands-on experiments, and dramatic presentation.	You might administer a pretest and develop learning contracts based on the results. Some students might have to complete activities to address foundational knowledge they do not demonstrate, others can complete the activities as outlined in the lesson, and yet others might be ready for extension activities. If no teaching aides are available, a more advanced student might facilitate the activities as outlined in the lesson while you work with other students.
Time	No adjustments are made in the plan.	Allow extra time at the end of each lesson for students who require extra instructional time. Provide enrichment experiences for those who already demonstrate mastery.
Space	Though various grouping configurations are used, no specific modifications for use of space are evident in the plan.	Work with students to quickly shift furniture from whole-class to small-group setups.

Differentiating Product

Remember that every student is responsible for demonstrating understanding as outlined in the criteria in Table 6.1. But you can differentiate the formative and summative assessments used to measure understanding. The lessons in this chapter lend themselves to tiered tasks. Refer to Table 3.3, page 139, for specific suggestions to tier the inquiry tasks. Two of the suggestions fit these lessons particularly well:

- Provide a greater choice of materials for some students but a more limited choice to guide the inquiry of other students.
- Provide formats for tables and charts for the students who might need them.

CHAPTER 6: Understanding Changes in Motion

Ties to Literacy and Numeracy (Mathematics)

These lessons tie nicely to mathematics abilities and understandings, including measurement of time and distance, making predictions and testing those predictions, gathering data, and using data to formulate explanations. Connections to the National Council of Teachers of Mathematics (NCTM 2000) standards are summarized in Table 6.6 and connections to the *Common Core State Standards for Mathematics* (2010) are found in Table 6.7, page 229.

Table 6.6

Mathematics Standards Supported by Science Instruction

Mathematics Standard	Expectations of the Mathematics Standard	Science Connections
Numbers and Operations: Understand numbers, ways of representing numbers, relationships among numbers, and number systems	Understand the place-value structure of the base-ten number system and be able to represent and compare whole numbers and decimals	Students use a meterstick and must report measurements using whole numbers and decimals.
Algebra: Understand patterns, relations, and functions	• Describe, extend, and make generalizations about geometric and numeric patterns • Represent and analyze patterns and functions using words, tables, and graphs	These standards are addressed in the various mathematical representations of experimental results.
Algebra: Use mathematical models to represent and understand quantitative relationships	Model problem situations with objects and use representations such as graphs, tables, and equations to draw conclusions	
Algebra: Analyze change in various contexts	• Investigate how change in one variable relates to change in a second variable • Identify and describe situations with constant or varying rates of change and compare them	
Measurement: Apply appropriate techniques, tools, and formulas to determine measurements	Select and apply appropriate standard units and tools to measure length, area, volume, weight, time, temperature, and the size of angles	This is an important ability to use during the various experiments.

(continued)

Table 6.6 *(continued)*

Mathematics Standard	Expectations of the Mathematics Standard	Science Connections
Data Analysis/ Probability: Formulate questions that can be addressed with data and collect, organize, and display relevant data to answer them	• Collect data using observations, surveys, and experiments • Represent data using tables and graphs such as line plots, bar graphs, and line graphs • Recognize the differences in representing categorical and numerical data	These are also important abilities for the various investigations.
Data Analysis/ Probability: Develop and evaluate inferences and predictions that are based on data	Propose and justify conclusions and predictions that are based on data and design studies to further investigate the conclusions or predictions	The importance of these abilities in this lesson is found not only when students give priority to evidence in responding to questions and formulate explanations from evidence, but also when they communicate and justify explanations.
Data Analysis/ Probability: Understand and apply basic concepts of probability	Predict the probability of outcomes of simple experiments and test the predictions	In Lesson #1, students are asked to predict which marble will reach the bottom of the ramp first and then to design their own experiments.

Table 6.7

Common Core State Standards for Mathematics (Grades 3–5) With Connections to Science

	Summary of Common Core State Standards	**Science Connections to Lessons #1–4 in Chapter 6**
3rd Grade	Represent and Interpret Data–Draw scaled picture graphs and bar graphs to represent data sets; generate measurement data and show the data by making a line plot	This standard ties nicely to any experimental work in the lessons in this chapter.
	Problem-solving (included across the standards) related to measurement and estimation of time intervals, liquid volumes, and masses of objects	Time intervals are measured in the experiments.
4th Grade	Represent and Interpret Data—Make a line plot to display a set of measurements in fractions of a unit	This standard also relates in multiple ways to experiments the students do.
	Problem-solving (included across the standards) related to measurement and estimation of time intervals, liquid volumes, and masses of objects	
5th Grade	Understand place values, including writing decimals to the thousandths place	Students must understand place value when using the meterstick.
	Represent and Interpret Data—Make a line plot to display a data set of measurements in fractions of a unit	This standard also relates to experiments the students are required to complete.

Literacy connections are also easy to find in these lessons and range from using vocabulary (using words to describe motion, connecting this language to real motion, and practicing these words in a real-world context) to sharing ideas, through both writing and speaking. You can build connections to the reading standards for informational text, focusing on craft and structure (meanings of domain-specific words). Writing can be emphasized by having students write explanations (i.e., informative or explanatory text to examine and convey ideas). Speaking and listening are used extensively in these lessons, and a key standard is comprehension and collaboration (engaging in collaborative discussions). These connections apply to each of the grade levels (third through fifth).

CHAPTER 6: Understanding Changes in Motion

Consideration Across the Grades

As noted earlier, after the direct kinesthetic experiences, students should be encouraged to represent their ideas with words and on paper. With students in fourth grade or above, these representations can include mathematical models and motion graphs. In general, mathematical models and graphs should be more extensively incorporated for fourth and fifth graders.

In particular, Lesson #4 asks that students provide evidence for their understandings. Fourth- or fifth-grade students can be pressed to present quantitative evidence to support their results. Students should create their own methods of using a stopwatch or a meterstick to show that the time it takes for the ball to roll the first half of the motion is the same as the time it takes for the ball to roll the second half of the motion (one criterion for knowing that the speed is not changing).

Build Your Library

For Your Students
Books
Experiments With Friction (Salvatore Tocci)
Eyewitness: Force and Motion (Peter Lafferty)
Forces and Motion (Lisa Trumbauer)
Forces and Movement (Straightforward Science) (Peter D. Riley)
Go, Go, Go! Kids on the Move (Stephen R. Swinburne)
May the Force Be With You! (Marsha Riddle Buly and Nicole Melvin)
The Magic School Bus Plays Ball: A Book About Forces (Joanna Cole)
Start and Stop (The Way Things Move) (Lola M. Schaefer)

For You
Books
Uncovering Student Ideas in Physical Science, Vol. I: 45 New Force and Motion Assessment Probes (Keeley and Harrington)
Force and Motion: Stop Faking It! Finally Understanding Science So You Can Teach It (Robertson)

Online
Force and Motion (SciPack in the NSTA Learning Center)
Force and Motion: Stop Faking It! (Two archived NSTA Web Seminars in the NSTA Learning Center)

Topic: Forces (K–4)
Go to: *www.scilinks.org*
Code: HTT013

Topic: Force and Motion (K–4)
Go to: *www.scilinks.org*
Code: HTT014

Topic: Force and Motion: Newton's First Law (5–8)
Go to: *www.scilinks.org*
Code: HTT015

Topic: Force and Motion: Position and Motion (5–8)
Go to: *www.scilinks.org*
Code: HTT016

Appendixes

Appendix A

Planning Template for the Predictive Phase

	Lesson Topic	
Essential Understandings		
Knowledge Required from Previous Instruction		
Knowledge and Skills to Be Learned	**Concepts:** See the Learning Targets. **Vocabulary:** **Skills:**	
Criteria to Demonstrate Understanding		
Learning Targets and Instructional Plans		
Target #1		
	Instructional Plan for Learning Target #1: Will be completed during the responsive phase.	
Target #2		
	Instructional Plan for Learning Target #2: Will be completed during the responsive phase.	
Target #3		
	Instructional Plan for Learning Target #3: Will be completed during the responsive phase.	
Target #4		
	Instructional Plan for Learning Target #4: Will be completed during the responsive phase.	

Hard-to-Teach Science Concepts

Appendixes

Appendix B

Strategy Selection Template

	Learning Target #1	Learning Target #2	Learning Target #3	Learning Target #4
Possible Strategies for:				
Identifying Preconceptions				
Eliciting and Confronting Preconceptions				
Sense Making				
Demonstrating Understanding				
Selected Inquiry Strategy				
Selected Metacognitive Strategy				

Appendix C

Planning Template for the Responsive Phase*

Identifying Student Preconceptions	

Learning Target #1:
Instructional Plan

Elicit Preconceptions:

Confront Preconceptions:

Use Sense-Making Experiences:

Use Formative Assessments:

Learning Target #2:
Instructional Plan

Elicit Preconceptions:

Confront Preconceptions:

Use Sense-Making Experiences:

Use Formative Assessments:

* The instructional plans for each learning target as you write them will not account for changes that will come about during instruction or re-teaching.

(continued)

Planning Template for the Responsive Phase *(continued)*

Learning Target #3:
Instructional Plan
Elicit Preconceptions: Confront Preconceptions: Use Sense-Making Experiences: Use Formative Assessments:

Learning Target #4:
Instructional Plan
Elicit Preconceptions: Confront Preconceptions: Use Sense-Making Experiences: Use Formative Assessments:

References

Agan, L., and C. Sneider. 2004. Learning about the Earth's shape and gravity: A guide for teachers and curriculum developers. *Astronomy Education Review* 2 (2): 90–117. Online at *http://aer.noao.edu*.

Ambron, J. 1987. Writing to improve learning in biology. *Journal of College Science Teaching* 16 (4): 263–266.

American Association for the Advancement of Science (AAAS). 1993. *Benchmarks for science literacy*. New York: Oxford University Press.

American Association for the Advancement of Science (AAAS). 2001a. *Designs for science literacy*. New York: Oxford University Press.

American Association for the Advancement of Science (AAAS). 2001b. *Atlas of science literacy*. Vol. 1. Washington DC: AAAS.

American Association for Advancement of Science (AAAS). 2007. *Atlas of science literacy*. Vol. 2. Washington DC: AAAS.

American Association for the Advancement of Science (AAAS). 2009. Benchmarks for science literacy online: *www.project2061.org/publications/bsl/online*

Andersson, B. 1990. Pupils' conceptions of matter and its transformation (age 12–16). *Studies in Science Education* 18 (1): 53–85.

Aristotle. c. 350 BC/1971. *De Caelo* or *De Caelo et Mundo*. (On the Heavens.) English translation. Loeb Classic Greek Library. Cambridge, MA: Harvard University Press.

Arons, A. 1983. Student patterns of thinking and reasoning, part one. *The Physics Teacher* 21(12): 576–581.

Arons, A. 1984. Student patterns of thinking and reasoning, part two. *The Physics Teacher* 22 (1): 21–26.

ASCD. 2008. Analyzing classroom discourse to advance teaching and learning. *Education Update* 50 (2): 1, 3, and 7.

Atkin, J. M., and J. E. Coffey, eds. 2003. *Everyday assessment in the science classroom*. Arlington, VA: NSTA Press.

Atkinson, H., and S. Bannister. 1998. Concept maps and annotated drawings: A comparative study of two assessment tools. *Primary Science Review* 51: 3–5.

Baker, L. 2004. Reading comprehension and science inquiry: Metacognitive connections. In *Crossing borders in literacy and science instruction*, ed. E. W. Saul, 239–257. Newark, DE: International Reading Association and Arlington, VA: NSTA Press.

Bar, V. 1989. Children's views about the water cycle. *Science Education* 73 (4): 481–500.

Bar, V., C. Sneider, and N. Martimbeau. 1997. Is there gravity in space? *Science and Children* (Apr.): 38–43.

Barton, M. L., and D. L. Jordan. 2001. *Teaching reading in science: A supplement to teaching reading in the content areas teacher's manual*. 2nd ed. Aurora, CO: McREL.

Beals, K., and C. Willard. 2007. *Space science sequence for grades 3–5: Teacher's guide*. GEMS Sequences. Berkeley, CA: Lawrence Hall of Science, University of California.

Bentley, M. L., E. S. Ebert II, and C. Ebert. 2007. *Teaching constructivist science: Nurturing natural investigators in the standards-based classroom*. Thousand Oaks, CA: Corwin Press.

Black, P. J., and C. Harrison. 2004. *Science inside the black box: Assessment for learning in the science classroom*. London: nferNelson Publishing Company.

References

Black, P., C. Harrison, C. Lee, B. Marshall, and D. Wiliam. 2003. *Assessment for learning: Putting it into practice.* London: Open University Press.

Blakey, E., and S. Spence. 1990. Developing metacognition. (ERIC Digest no. ED327218.) *www.ericdigests.org/pre-9218/developing.htm*

Bransford, J., A. Brown, and R. R. Cocking. 1999. *How people learn: Brain, mind, experience, and school.* Washington, DC: National Academies Press.

Brown, D. S. 2003. High school biology: A group approach to concept mapping. *The American Biology Teacher* 65 (3): 192–197.

Bruner, J., and H. Haste, eds. 1987. *Making sense: The child's construction of the world.* London: Methuen.

Burke, K. A., T. J. Greenbowe, and B. M. Hand. 2005. The process of the science writing heuristic. Iowa State University. *http://avogadro.chem.iastate.edu/SWH/homepage.htm*

Burke, K. A., B. Hand, J. Poock, and T. Greenbowe. 2005. Using the science writing heuristic: Training chemistry teaching assistants. *Journal of College Science Teaching* 35 (1): 36–41.

Bybee, R. W. 1997. *Achieving scientific literacy: From purposes to practices.* Portsmouth, NH: Heinemann.

Bybee, R. W. 2006. How inquiry can contribute to the prepared mind. *The American Biology Teacher* 68 (8): 454–457.

Campbell, B., and L. Fulton. 2003. *Science notebooks: Writing about inquiry.* Portsmouth, NH: Heinemann.

Cañas, A. J., and J. D. Novak. 2006. Re-examining the foundations for effective use of concept maps. Cmap Tools. *http://cmc.ihmc.us/cmc2006Papers/cmc2006-p247.pdf*

Carlsen, W. H. 2007. Language and science learning. In *Handbook of research on science education*, ed. S. K. Abell and N. G. Lederman, 57–74. Hillsdale, NJ: Lawrence Erlbaum.

Cartier, J. 2000. Research report: Using a modeling approach to explore scientific epistemology with high school biology students. Wisconsin Center for Education Research. *www.wcer.wisc.edu/NCISLA/publications/reports/RR991.pdf*

Carey, S. 1985. *Conceptual change in childhood.* Cambridge, MA: MIT Press.

Cawelti, G. C. 1999. *Handbook of research on improving student achievement.* 2nd ed. Arlington, VA: Educational Research Service.

Century, J. R., J. Flynn, D. S. Makang, M. Pasquale, K. M. Roblee, J. Winokur, and K. Worth. 2002. Supporting the science-literacy connection. In *Learning science and the science of learning*, ed. R. W. Bybee, 37–49. Arlington, VA: NSTA Press.

Champagne, A., L. Klopfer, and J. Anderson. 1980. Factors influencing the learning of classical mechanics. *American Journal of Physics* 48 (12): 1074.

Children's Learning in Science Project (CLIS). 1987. *CLIS in the classroom: Approaches to teaching.* Leeds, UK: University of Leeds, Children's Learning in Science Project, Centre for Studies in Science and Mathematics Education.

Chin, C., D. E. Brown, and C. B. Bertram. 2002. Student-generated questions: A meaningful aspect of learning in science. *International Journal of Science Education* 24 (5): 521–549.

Coffey, J. E., and J. M. Atkin, eds. 2003. *Everyday assessment in the science classroom.* Arlington, VA: NSTA Press.

Colbert, J. T., J. K. Olson, and M. P. Clough. 2007. Using the web to encourage student-generated questions in large-format introductory biology classes. *CBE Life Sciences Education* 6 (1): 42–48.

Common Core State Standards Initiative. 2010. The Standards. Retrieved June 5, 2010, from the Common Core State Standards Initiative: Preparing America's Students for College & Career website: *www.corestandards.org/the-standards*

Costa, A. 2008. The thought-filled curriculum. *Educational Leadership* 65 (5): 20–24.

Costa, A., and B. Kallick, eds. 2000. *Discovering and exploring habits of mind.* Alexandria, VA: ASCD.

Crowther, D., and J. Cannon. 2004. Strategy makeover: K-W-L to T-H-C. *Science and Children* 42 (1): 42–44.

Davies, A. 2003. Learning through assessment: Assessment for learning in the science classroom. In *Everyday assessment in the science classroom,* ed. J. M. Atkin and J. E. Coffey, 13–25. Arlington, VA: NSTA Press.

Dewey, J. 1902. *The child and the curriculum.* Chicago: University of Chicago Press.

Donovan, M. S., and J. D. Bransford. 2005. *How students learn: Science in the classroom.* Washington, DC: National Academies Press.

Donovan, M. S., J. D. Bransford, and J. W. Pellegrino. 1999. *How people learn: Bridging research and practice.* Washington, DC: National Academies Press.

Driver, R., A. Squires, P. Rushworth, and V. Wood-Robinson. 1994. *Making sense of secondary science: Research into children's ideas.* London: Routledge.

Duschl, R.A., H.A. Schweingruber, and A.W. Shouse. 2007. *Taking science to school: Learning and teaching science in grades k–8.* Washington, DC: National Academies Press.

Dykstra, D., and D. Sweet. 2009. Conceptual development about motion and force in elementary and middle school students. *American Journal of Physics* 77 (5): 468–476.

Edens, K. M., and E. Potter. 2003. Using descriptive drawings as a conceptual change strategy in elementary science. *School Science and Mathematics* 103 (3): 135–44.

Eichinger, J. 2009a. *Activities linking science with math, K–4.* Arlington, VA: NSTA Press.

Eichinger, J. 2009b. *Activities linking science with math, 5–8.* Arlington, VA: NSTA Press.

Ekici, F., E. Ekici, and F. Aydin. 2007. Utility of concept cartoons in diagnosing and overcoming misconceptions related to photosynthesis. *International Journal of Environmental & Science Education* 2 (4): 111–124.

Enger, S. K., and R. E. Yager. 2009. *Assessing student understanding in science.* 2nd ed. Thousand Oaks, CA: Corwin Press.

Eyon, B. S., and M. Linn. 1988. Learning and instruction: An examination of four research perspectives in science education. *Review of Educational Research* 58 (3): 251–301.

Fisher, K., J. H. Wandersee, and D. E. Moody. 2000. *Mapping biology knowledge.* Boston: Kluwer.

Freedman, R. L. H. 1994. *Open-ended questioning: A handbook for educators.* Menlo Park, CA: Addison-Wesley.

Fulwiler, B. 2007. *Writing in science: How to scaffold instruction to support learning.* Portsmouth, NH: Heinemann.

Gardner, H. 1983/2004. *Frames of Mind: The theory of multiple intelligences.* New York: Basic Books.

References

Gilbert, S. W., and S. W. Ireton. 2003. *Understanding models in Earth and space science*. Arlington, VA: NSTA Press.

Gilsdorf, D., and P. Loftus. Personal communication, September 15, 2010.

Given, B. K. 2002. *Teaching to the brain's natural learning systems*. Alexandria, VA: ASCD.

Goldston, M. J., ed. 2004. *Stepping up to science and math: Exploring the natural connections*. Arlington, VA: NSTA Press.

Gore, M. C. 2004. *Successful inclusion strategies for secondary and middle school teachers: Keys to help struggling learners access the curriculum*. Thousand Oaks, CA: Corwin Press.

Gregory, G. H., and E. Hammerman. 2008. *Differentiated instructional strategies for science, grades K–8*. Thousand Oaks, CA: Corwin Press.

Grosslight, L., C. Unger, E. Jay, and C. L. Smith. 1991. Understanding models and their use in science—Conceptions of middle and high school students and experts. *Journal of Research in Science Teaching* 28 (9): 799–822.

Hake, R. 1998. Interactive-engagement versus traditional methods: A six-thousand-student survey of mechanics test data for introductory physics courses. *American Journal of Physics* 66 (1): 64–74.

Hale, M. S., and E. A. City. 2006. *The teacher's guide to leading student-centered discussions: Talking about texts in the classroom*. Thousand Oaks, CA: Corwin Press.

Hall, T., N. Strangman, and A. Meyer. 2009. Differentiated instruction and implications for UDL implementation. *www.cast.org/publications/ncac/ncac_diffinstructudl.html*

Hammerman, E., and D. Musial. 2008. *Integrating science with mathematics and literacy: New visions for learning and assessment*. Thousand Oaks, CA: Corwin Press.

Hand, B., L. Hockenberry, K. Wise, and L. Norton-Meier, eds. 2008. *Questions, claims and evidence: The important place of argument in children's science writing*. Portsmouth, NH: Heinemann.

Hargrove, T. Y., and C. Nesbit. 2003. *Science notebooks: Tools for increasing achievement across the curriculum*. Columbus, OH: ERIC Clearinghouse for Science, Mathematics, and Environmental Education. ERIC Digest no. ED482720. *www.ericdigests.org/2004-4/notebooks.htm*

Harlen, W. 2000. *Teaching, learning & assessing science 5–12*. 3rd ed. London: Paul Chapman Publishing.

Harlen, W. 2001. *Primary science: Taking the plunge*. 2nd ed. Portsmouth, NH: Heinemann Educational Books.

Harrison, A. G., and R. K. Coll. 2008. *Using analogies in middle and secondary science classrooms: The FAR guide—An interesting way to teach with analogies*. Thousand Oaks, CA: Corwin Press.

Hartman, H. J., and N. A. Glasgow. 2002. *Tips for the science teacher: Research-based strategies to help students learn*. Thousand Oaks, CA: Corwin Press.

Harvey, S., and A. Goudvis. 2007. *Strategies that work: Teaching comprehension for understanding and engagement*. 2nd ed. Portland, ME: Stenhouse.

Harvey, S., and A. Goudvis. 2011. *The comprehension toolkit: Language and lessons for active literacy*. Portsmouth, NH: Heinemann.

Haury, D. K. 2001. *Teaching science through inquiry with archived data*. Columbus, OH: ERIC Clearinghouse for Science, Mathematics, and Enviornmental Education. ERIC Digest no. ED465545. *www.eric.ed.gov/ERICWebPortal/detail?accno=ED465545*

Hawking, S. W. 1988. *A brief history of time.* New York: Bantam Books.

Hazen, R. M., and J. Trefil. 2009. *Science matters: Achieving scientific literacy.* New York: Anchor Books.

Heritage, M. 2008. Learning progressions: Supporting instruction and formative assessment. Paper prepared for the Formative Assessment for Teachers and Students (FAST) Council of Chief State School Officers, Washington, DC.

Hipkins, R., R. Bolstad, R. Baker, A. Jones, M. Barker, B. Bell, R. Coll, B. Cooper, M. Forret, A. Harlow, I. Taylor, B. France, and M. Haigh. 2002. *Curriculum, learning and effective pedagogy: A literature review in science education.* Retrieved March 23, 2008, from the New Zealand Council for Educational Research website: www.nzcer.org.nz/default.php?products_id=559

Horton, P., A. McConney, M. Gallo, A. Woods, G. Senn, and D. Hamelin. 1993. An investigation of the effectiveness of concept mapping as an instructional tool. *Science Education* 77 (1): 95–111.

Hyerle, D. 2000. *A field guide to using visual tools.* Alexandria, VA: ASCD.

Ingram, M. 1993. *Bottle biology: An idea book for exploring the world through soda bottles and other recyclable materials.* Dubuque, IA: Kendall/Hunt.

Inhelder, B., and J. Piaget. 1958. *The growth of logical thinking from childhood to adolescence.* London: Routledge and Kegan Paul.

International Centre for Development Oriented Research in Agriculture (ICRA). n.d. Systems diagrams: Guidelines. ICRA. www.icra-edu.org/objects/anglolearn/Systems_Diagrams-Guidelines1.pdf

Jensen, E. 1998. *Teaching with the brain in mind.* Alexandria, VA: ASCD.

Jones, D. J. 2007. The station approach: How to teach with limited resources. *Science Scope* 30 (6): 16–27.

Keeley, P. 2005. *Science curriculum topic study: Bridging the gap between standards and practice.* Thousand Oaks, CA: Corwin Press and Arlington, VA: NSTA Press.

Keeley, P. 2008. *Science formative assessment: 75 practical strategies for linking assessment, instruction, and learning.* Thousand Oaks, CA: Corwin Press and Arlington, VA: NSTA Press.

Keeley, P. 2011. *Uncovering student ideas in life science, vol. 1: 25 new formative assessment probes.* Arlington, VA: NSTA Press.

Keeley, P., and F. Eberle. 2008. Using standards and cognitive research to inform the design and use of formative assessment probes. In *Assessing science learning,* ed. J. Coffey, R. Douglas, and C. Stearns, 206–207. Arlington, VA: NSTA Press.

Keeley, P., F. Eberle, and C. Dorsey. 2008. *Uncovering student ideas in science, vol. 3: Another 25 formative assessment probes.* Arlington, VA: NSTA Press.

Keeley, P., F. Eberle, and L. Farrin. 2005. *Uncovering student ideas in science, vol. 1: 25 formative assessment probes.* Arlington, VA: NSTA Press.

Keeley, P., F. Eberle, and J. Tugel. 2007. *Uncovering student ideas in science, vol. 2: 25 more formative assessment probes.* Arlington, VA: NSTA Press.

Keeley, P., and R. Harrington. 2010. *Uncovering student ideas in physical science, vol. 1: 45 new force and motion assessment probes.* Arlington, VA: NSTA Press.

Keeley, P., and C. M. Rose. 2006. *Mathematics curriculum topic study: Bridging the gap between standards and practice.* Thousand Oaks, CA: Corwin Press.

References

Keeley, P., and J. Tugel. 2009. *Uncovering student ideas in science, vol. 4: 25 new formative assessment probes.* Arlington, VA: NSTA Press.

Keogh, B., and S. Naylor. 1999. Concept cartoons, teaching and learning in science: An evaluation. *International Journal of Science Education* 21 (4): 431–446.

Keogh, B., and S. Naylor. 2007. Talking and thinking in science. *School Science Review* 88 (324): 85–90.

Koba, S., with A. Tweed. 2009. *Hard-to-teach biology concepts: A framework to deepen student understanding.* Arlington, VA: NSTA Press.

Konicek-Moran, R. 2008. *Everyday science mysteries: Stories for inquiry-based science teaching.* Arlington, VA: NSTA Press.

Konicek-Moran, R. 2009. *More everyday science mysteries: Stories for inquiry-based science teaching.* Arlington, VA: NSTA Press.

Konicek-Moran, R. 2010. *Even more everyday science mysteries: Stories for inquiry-based science teaching.* Arlington, VA: NSTA Press.

Konicek-Moran, R. 2011. *Yet more everyday science mysteries: Stories for inquiry-based science teaching.* Arlington, VA: NSTA Press.

Köse, S. 2008. Diagnosing student misconceptions: Using drawings as a research method. *World Applied Sciences Journal* 3 (2): 283–293.

Krajcik, J., P. C. Blumenfeld, R. W. Marx, K. M. Bass, and J. Fredricks. 1998. Inquiry in project-based science classrooms: Initial attempts by middle school students. *The Journal of the Learning Sciences* 7 (3 and 4): 313–350.

Krueger, A., and J. Sutton. 2001. *EDThoughts: What we know about science teaching and learning.* Aurora, CO: McREL.

Lawrence Hall of Science, University of California, Berkeley. 2010. Seeds of science/Roots of reading website. *http://seedsofscience.org*

Layman, J., G. Ochoa, and H. Heikkinen. 1996. *Inquiry and learning: Realizing science standards in the classroom.* New York: National Center for Cross-Disciplinary Teaching and Learning.

Lazear, D. 1991. *Seven ways of teaching: The artistry of teaching with multiple intelligences.* Palatine, IL: IRI/Skylight Publishing.

Leach, J., R. Driver, P. Scott, and C. Wood-Robinson. 1992. *Progression in conceptual understanding of ecological concepts by pupils aged 5–16.* Leeds, UK: Centre for Studies in Science and Mathematics Education, University of Leeds.

Leeds National Curriculum Science Support Project (LNCSSP). 1992a. *Children's ideas about ecosystems: Research summary. www.learner.org/courses/essential/life/support/2_Ecosystems.pdf*

Leeds National Curriculum Science Support Project (LNCSSP). 1992b. *Children's ideas about nutrition: Research summary. www.learner.org/courses/essential/life/support/3_Nutrition.pdf*

Leo, O., D. C. Eichinger, C. W. Anderson, G. D. Berkhamer, and T. S. Blakeslee. 1993. Changing middle school students' conceptions of matter and molecules. *Journal of Research in Science Teaching* 30 (3): 249–270.

Lipton, L., and B. Wellman. 1998. *Pathways to understanding: Patterns and practices in the learning-focused classroom.* 3rd ed. Sherman, CT: MiraVia.

Llewellyn, D. 2007. *Inquire within: Implementing inquiry-based science standards in grades 3–8.* Thousand Oaks, CA: Corwin Press.

Lochhead, J. Personal communication, September 22, 2010.

Lowery, L. F. 1990. *The biological basis of thinking and learning.* Berkeley, CA: University of California Press.

Luft, J., R. L. Bell, and J. Gess-Newsome, eds. 2008. *Science as inquiry in the secondary setting.* Arlington, VA: NSTA Press.

Magnusson, S. J., and A. S. Palinscar. 2004. Learning from text designed to model scientific thinking in inquiry-based instruction. In *Crossing borders in literacy and science instruction,* ed. E. W. Saul, 316–339. Newark, DE: International Reading Association and Arlington, VA: NSTA Press.

Mali, G. B., and A. Howe. 1979. Development of Earth and gravity concepts among Nepali children. *Science Education* 63 (5): 685–691.

Martin, P. M. 2006. *Elementary science methods: A constructivist approach.* 4th ed. Belmont, CA: Thomson/Wadsworth.

Marzano, R. J. 1992. *A different kind of classroom: Teaching with dimensions of learning.* Alexandria, VA: ASCD.

Marzano, R. J., and M. J. Haystead. 2008. *Making standards useful in the classroom.* Alexandria, VA: ASCD.

Marzano, R. J., D. J. Pickering, and J. E. Pollock. 2001. *Classroom instruction that works: Research-based strategies for increasing student achievement.* Alexandria, VA: ASCD.

Masilla, V. B., and H. Gardner. 2008. Disciplining the mind. *Educational Leadership* 65 (5): 14–19.

McConnell, S. 1993. Talking drawings: A strategy for assisting learners. *Journal of Reading* 36 (4): 260–269.

McDermott, L., and M. Rosenquist. 1987. A conceptual approach to teaching kinematics. *American Journal of Physics* 55 (5): 407–415.

McDermott, L. and the Physics Education Group at the University of Washington. 1996. *Physics by inquiry,* Vols. 1 and 2. New York: John Wiley & Sons.

McMahon, M., P. Simmons, R. Sommers, D. DeBaets, and F. Crawley, eds. 2006. *Assessment in science: Practical experiences and education research.* Arlington, VA: NSTA Press.

McNair, S., and M. Stein. 2001. Drawing on their understanding: Using illustrations to invoke deeper thinking about plants. Paper presented at the annual meeting of the Association for the Education of Teachers of Science, Costa Mesa, CA.

McNeill, K. L., and J. Krajcik. 2006. Supporting students' construction of scientific explanation through generic versus context-specific written scaffolds. Paper presented at the annual meeting of the American Educational Research Association, San Francisco, CA.

McNeill, K. L., and J. Krajcik. 2008. Inquiry and scientific explanations: Helping students use evidence and reasoning. In *Science as inquiry in the secondary settings,* ed. J. Luft, R. L. Bell, and J. Gess-Newsome, 121–133. Arlington, VA: NSTA Press.

McTighe, J., and G. Wiggins. 1999. *The understanding by design handbook.* Alexandria, VA: ASCD.

McTighe, J., and G. Wiggins. 2004. *Understanding by design professional development workbook.* Alexandria, VA: ASCD.

Michaels, S., A. W. Shouse, and H. A. Schweingruber. 2008. *Ready, set, science! Putting research to work in K–8 science classrooms.* Washington, DC: National Academies Press.

Mind Tools. n.d. Systems diagrams. *www.mindtools.com/pages/article/newTMC_04.htm*

References

Minstrell, J. 1989. Teaching science for understanding. In *Toward the thinking curriculum: Current cognitive research,* ed. L. B. Resnick and L. E. Klopfer, 129–149. Alexandria, VA: ASCD.

Minstrell, J. 2000. Facet-based instruction. *www.diagnoser.com*

Minstrell, J., R. Anderson, P. Kraus and J. E. Minstrell. 2008. From practice to research and back: Perspectives and tools in assessing for learning. In *Science assessment: Research and practical approaches,* ed. J. Coffey, R. Douglas and C. Stearns, 37–67. Arlington, VA: NSTA Press.

Mohan, L., and C. W. Anderson. 2009. Teaching experiments and the carbon cycle learning progression. Paper presented at the Learning Progressions in Science (LeaPS) Conference, Iowa City, IA.

Mortimer, E. F., and P. H. Scott. 2003. *Meaning making in secondary science classrooms.* London: Open University Press.

National Council of Teachers of Mathematics (NCTM). 2000. *Principles and standards for school mathematics.* Reston, VA: NCTM.

National Research Council (NRC). 1996. *National science education standards.* Washington, DC: National Academies Press.

National Research Council (NRC). 2000. *Inquiry and the national science education standards: A guide for teaching and learning.* Washington, DC: National Academies Press.

National Research Council (NRC). 2001. *Classroom assessment and the national science education standards.* Washington, DC: National Academies Press.

National Research Council (NRC). 2005. *How students learn: Science in the classroom.* Committee on *How People Learn,* A Targeted Report for Teachers, ed. M. S. Donovan and J. D. Bransford, Division of Behavioral and Social Sciences and Education. Washington, DC: National Academies Press.

National Research Council (NRC). 2010. Conceptual framework for the next generation science education standards. Washington, DC: National Academies Press.

National Science Teachers Association (NSTA). 2004. NSTA position statement: Students with disabilities. *www.nsta.org/about/positions/disabilities.aspx*

Naylor, S., and B. Keogh. 2000a. *Concept cartoons in science education.* London: Millgate House Education.

Naylor, S., and B. Keogh. 2000b. *Concept cartoons in science education* (CD). London: Millgate House Education.

Nelson-Herber, J. 1986. Expanding and refining vocabulary in content areas. *Journal of Reading* 29 (7): 626–33.

Newton, I. 1687/1971. *Philosophiae Naturalis Principia Mathematica.* 3rd ed. (1726) with variant readings. Assembled and edited by Alexandre Koyré and I. Bernard Cohen with the assistance of Anne Whitman. Cambridge, MA: Harvard University Press.

Novak, J. D. 1996. Concept mapping: A tool for improving science teaching and learning. In *Improving teaching and learning in science and mathematics,* ed. D. F. Treagust, R. Duit, and B. J. Fraser, 32–43. New York: Teachers College Press.

Novak, J. D. 1998. *Learning, creating, and using knowledge: Concept maps as facilitative tools in schools and corporations.* Hillsdale, NJ: Lawrence Erlbaum.

Nussbaum, J., and J. D. Novak. 1976. An assessment of children's concepts of the Earth utilizing structured interviews. *Science Education* 60 (4): 535–550.

Nussbaum, J. 1979. Children's conceptions of the Earth as a cosmic body: A cross age study. *Science Education* 63 (1): 83–93.

O'Neill, G., and T. McMahon. 2005. Student-centered learning: What does it mean for students and lecturers? Ireland Society for Higher Education (ASHE) Readings website: *www.aishe.org/readings/2005-1/index.html*

Palinscar, A. S., and A. L. Brown. 1985. Reciprocal teaching: Activities to promote "reading with your mind." In *Reading, thinking, and concept development*, ed. T. L. Harris and E. J. Cooper, 147–158. New York: College Board Publications.

Piaget, J. 1929. *The child's conception of the world.* London: K. Paul, Trench, Trubner & Co.

Piaget, J., and B. Inhelder. 1956. *The child's conception of space*. London: Routledge.

Posner, G., K. Strike, P. Hewson, and W. Gertzog. 1982. Accommodation of a scientific conception: Toward a theory of conceptual change. *Science Education* 66 (2): 211–227.

Purvis, D. 2006. Fun with phase changes. *Science and Children* 43 (5): 23–25.

Rangahau, W. M. 2002. *Curriculum, learning, and effective pedagogy: A literature review in science education.* Wellington, New Zealand: Ministry of Education.

Rich, S. 2010. *Outdoor science: A practical guide.* Arlington, VA: NSTA Press.

Rico, G. 2000. *Writing the natural way.* New York: Putnam. Described in D. Hyerle, *A field guide to using visual tools*. Alexandria, VA: ASCD.

Ritchart, R., and D. Perkins. n.d. Visible thinking: Engaged students, in-depth learning, better teaching. Council for Exceptional Children. *www.cec.sped.org/AM/Template.cfm?Section+Home*

Ritchart, R., and D. Perkins. 2008. Making thinking visible. *Educational Leadership* 65 (5): 57–61.

Robertson, W. D. 2006. *Math: Stop faking it! Finally understanding science so you can teach it.* Arlington, VA: NSTA Press.

Rolheiser, C., and J. A. Ross. n.d. Student self-evaluation: What research says and what practice shows. Center for Developmental Learning. *www.cdl.org/resource-library/articles/self_eval.php?type_subject&id=4*

Rowell, P. 1997. Learning in school science: The promises and practices of writing. *Studies in Science Education* 30 (1): 19–56.

Russell, T., and D. Watt. 1990. *Evaporation and condensation.* SPACE Research Report. Liverpool, UK: Liverpool University Press.

Scott, P., H. Asoko, and J. Leach. 2007. Student conceptions and conceptual learning in science. In *Handbook of research on science education*, ed. S. K. Abell and N. G. Lederman, 31–55. Hillsdale, NJ: Lawrence Erlbaum.

Scott, V. G., and M. K. Weishaar. 2008. Talking drawings as a university classroom assessment technique. *The Journal of Effective Teaching* 8 (1): 42–51. *www.uncwil.edu/cte/ET/articles/Vol8_1/Scott.htm*

Sere, M. 1985. The gaseous state. In *Children's ideas in science*, ed. R. Driver, E. Guesne, and A. Tiberghein, 105–123. Milton Keynes, UK: Open University Press.

Shayer, M., and P. Adey, eds. 2002. *Learning intelligence.* London: Open University Press.

Smith, E., and C. Anderson. 1986. Alternative conceptions of matter cycling in ecosystems. Paper presented at the annual meeting of the National Association for Research in Science Teaching, San Francisco, CA.

References

Smutny, J. F., and S. E. Von Fremd. 2010. *Differentiating for the young child: Teaching strategies across the content areas, preK–3.* Thousand Oaks, CA: Corwin Press.

Sneider, C. 1986. *Earth, Moon, and stars: GEMS teacher's guide.* Berkeley, CA: Lawrence Hall of Science, University of California.

Sneider, C., and M. Ohadi. 1998. Unraveling students' misconceptions about the Earth's shape and gravity. *Science Education,* 82 (2): 265–284.

Sneider, C., and S. Pulos. 1983. Children's cosmographies: Understanding the Earth's shape and gravity. *Science Education* (67) 2: 205–221.

Sneider, C., S. Pulos, E. Freenor, J. Porter, and B. Templeton. 1986. Understanding the Earth's shape and gravity. *Learning '86* 14 (6): 43–47.

Southwest Center for Education and the Natural Environment. 2004. *The inquiry process.* SCENE. *http://scene.asu.edu/habitat/inquiry.html*

Stein, M., and S. McNair. 2002. Science drawings as a tool for analyzing conceptual understanding. Paper presented at the annual meeting of the Association for the Education of Teachers of Science, Charlotte, NC.

Stow, W. 1997. Concept mapping: A tool for self-assessment? *Primary Science Review* 49: 12–15.

Strike, K. A., and G. J. Posner. 1985. A conceptual change view of learning and understanding. In *Cognitive structure and conceptual change,* ed. L. West and R. Hamilton, 211–232. London: Academic Press.

Thier, M. 2002. *The new science literacy: Using language skills to help students learn science.* Portsmouth, NH: Heinemann.

Tomlinson, C. A. 1999. *The differentiated classroom: Responding to the needs of all learners.* Alexandria, VA: ASCD.

Tomlinson, C. A., S. N. Kaplan, J. S. Rensulli, J. Purcell, J. Leppien, and D. Burns. 2002. *The parallel curriculum: A design to develop high potential and challenge high-ability learners.* Thousand Oaks, CA: Corwin Press.

Tomlinson, C. A., and J. McTighe. 2006. *Integrating differentiated instruction and understanding by design.* Alexandria, VA: ASCD.

Trowbridge, D., and L. McDermott. 1981a. Investigation of student understanding of the concept of velocity in one dimension. *American Journal of Physics* 48 (12): 1020–1028.

Trowbridge, D., and L. McDermott. 1981b. Investigation of student understanding of acceleration in one dimension. *American Journal of Physics* 49 (2): 242–253.

Tweed, A. 2009. *Designing effective science instruction: What works in science classrooms.* Arlington, VA: NSTA Press.

Varelas, M., and C. C. Pappas. 2006. Intertextuality in read-alouds of integrated science-literacy units in primary classrooms: Opportunities for the development of thought and language. *Cognition and Instruction* 24 (2): 211–259.

Vasquez, J., M. W. Comer, and F. Troutman. 2010. *Developing visual literacy in science, K–8.* Arlington, VA: NSTA Press.

Vitale, M. R., and N. R. Romance. 2006. Research in science education: An interdisciplinary perspective. In *Teaching science in the 21st century,* ed. J. Rhoton and P. Shane, 329–351. Arlington, VA: NSTA Press.

Vosniadou, S., and W. F. Brewer. 1992. Mental models of the Earth: A study of conceptual change in childhood. *Cognitive Psychology* 24 (4): 535–585.

References

Vosniadou, S., and W. F. Brewer. 1994. Mental models of the day/night cycle. *Cognitive Science* 18 (1): 123–183.

Wallace, C. S., B. Hand, and E. Yang. 2004. The science writing heuristic: Using writing as a tool for learning in the laboratory. In *Crossing borders in literacy and science instruction*, ed. E. W. Saul, 355–367. Newark, DE: International Reading Association and Arlington, VA: NSTA Press.

Walsh, J. A., and B. D. Sattes. 2005. *Quality questioning: Research-based practice to engage every learner.* Thousand Oaks, CA: Corwin Press.

Wang, M. C., G. D. Haertel, and H. J. Walberg. 1993/1994. What helps students learn? *Educational Leadership* 51 (4) 74–79.

Weiss, I. R., J. D. Pasley, P. S. Smith, E. R. Banilower, and D. J. Heck. 2003. *Looking inside the classroom: A study of K–12 mathematics and science education in the United States.* Horizon Research International, Inside the Classroom. *www.horizon-research.com/insidetheclassroom/reports/looking*

Westcott, D. J., and D. L. Cunningham. 2005. Recognizing student misconceptions about science and evolution. *Mountain Rise Electronic Journal* 2 (2) (Spring/Summer). *http://facctr.wcu.edu/mountainrise/archive/vol2no2/html/science_evolution.html*

WestEd and Council of Chief State School Officers (CCSSO). 2007. Science assessment and item specifications for the 2009 National Assessment of Educational Progress. Prepublication edition. National Assessment Governing Board contract #ED04CO0148.

Wiggins, G., and J. McTighe. 1998. *Understanding by design.* Alexandria, VA: ASCD.

Wilhelm, J., and B. Edmiston. 1998. *Imagining to learn.* Portsmouth, NH: Heinemann.

Windschitl, M. 2008. What is inquiry? A framework for thinking about authentic scientific practice in the classroom. In *Science as inquiry in the secondary setting*, ed. J. Luft, R. L. Bell, and J. Gess-Newsome, 1–20. Arlington, VA: NSTA Press.

Wolfe, P. 2001. *Brain matters: Translating research into classroom practice.* Alexandria, VA: ASCD.

Woodruff, E., and K. Meyer. 1997. Explanations from intra- and inter-group discourse: Students building knowledge in the science classroom. *Research in Science Education* 27 (1): 25–39.

Wormeli, R. 2009. *Metaphors and analogies: Power tools for teaching any subject.* Portland, ME: Stenhouse Publishers.

Worth, K., J. Winokur, S. Crissman, M. Heller-Winokur, and M. Davis. 2009. *Science and literacy: A natural fit.* Portsmouth, NH: Heinemann.

Wright, A. W., and K. Bilica. 2007. Instructional tools to probe biology students' prior understanding. *The American Biology Teacher* 69 (1): 1–5.

Young, P. n.d. Visible thinking tools. San Diego State University's Encyclopedia of Educational Technology website: *http://edweb.sdsu.edu/eet/articles/VisThinkTools/start.htm*

Index

*Page numbers printed in **boldface** type refer to figures, tables, or Instructional Tools.*

A

A Different Kind of Classroom: Teaching with Dimensions of Learning, 33
A Field Guide to Using Visual Tools, **54, 117, 120**
Abstract concepts, 5, **5**
Abstract thinking, 208
"Accountable talk" skills, **103**
Activities Linking Science with Math, 5–8, **112**
Activities Linking Science with Math, K–4, **112,** 155
Adey, Phillip, 208
Adjustable assignments, **142**
Algebra, **146, 227**
Alternative conceptions of students, 15, 16. *See also* Misconceptions of students; Preconceptions of students
Analogies, **43, 109–110**
Anchor activities, 135, **135**
Annotated drawings, **44,** 47, **48,** 52, **56, 57, 121–122**
Anticipation guides, 47, 133, **138**
  for "The Flow of Matter and Energy in Ecosystems," 47, **49, 57**
Aristotle, **197**
Arons, Arnold, 206
Arts integration, 136
Assessing Student Understanding in Science, **26**
Assessment for Learning: Putting It Into Practice, **26**
Assessment in Science: Practical Experiences and Education Research, **26**
Assessments, 7
  differentiation of, 138, **139–140**
  formative, 7, **19,** 24, **26,** 42–44, 138
    determination of, 55, **55–56**
    for "Earth's Shape and Gravity," 192
    hints and resources for, **58**
    for "Matter and Its Transformations," 166, **167**
    probes, 91–92, 217–218
    for "The Flow of Matter and Energy in Ecosystems," 24, 55–56, **55–56, 58,** 73–74
    for "Understanding Changes in Motion," 217–218
  negotiated, 138, **139, 140**
  summative, 7, 24, **26,** 138
Astronomy curricula, 204
Astronomy lesson. *See* "Earth's Shape and Gravity"
Atlas of Science Literacy, **25, 26,** 28, 175

B

"Backwards design" process, 23–24, **26**
Benchmarks for Science Literacy, 20, **26,** 209
  related to "Earth's Shape and Gravity," 175, 177
  related to "Matter and Its Transformations," 161, **162**
  related to "The Flow of Matter and Energy in Ecosystems," 20, **21**
  related to "Understanding Changes in Motion," 211, 212–213
Big ideas, 7, 24
Bottle Biology TerrAqua Columns, **108**
Brainstorming webs, **44, 48, 114–115**
Brewer, William, 179, 180
Build your library
  astronomy, 204
  differentiation, 154–155
  force and motion, 230
  interdisciplinary approaches to teaching science, 155–156
  science trade books for students, 101, 154, 170, 204, 230
  states of matter, 170
Bybee, R. W., 18

C

Categorical organizers, **117**
Cause-and-effect organizers, **116**
CCM (Conceptual Change Model), 9, **10–11**
Centers, **135**
Choice boards, **134, 140**
  for "The Flow of Matter and Energy in Ecosystems," **141**
Choice in products or tasks, 138, **140**
Circle maps, **114–115**
Circle of Viewpoints, **84**
Claim/support/question, **83**
Classroom Instruction That Works: Research-based Strategies for Increasing Student Achievement, 54
Clock Buddies, **134**
Clustering, **114–115**
Cognitive challenges, 208
Cognitive learning, xvi, 30, 75
Comic Creator, **124**
ComicLife, **124**
Common Core State Standards Initiatives, 212
Comparison organizers, **117**
Complex concepts, 5, **5**
Computational Science Education Reference Desk, **78, 112**
Computer simulations, **113**
Concept cartoons, **44,** 52, **52,** 56, 57, **123–124**
  for "The Flow of Matter and Energy in Ecosystems," 68–69, **69**
Concept Cartoons in Science Education, **124**
Concept Cartoons website, **124**
Concept maps, **118–120,** 152
Conceptual Change Model (CCM), 9, **10–11**
Conceptual learning, 75
Concrete thinking, 208
Connecting Elementary Science and Literacy website, **99**
Connections, mathematical, **148**
Content Clips, **115, 120**
Content differentiation, 130–131
  for "Earth's Shape and Gravity," 200
  for "The Flow of Matter and Energy in Ecosystems," 130–131, **131**
  for "Understanding Changes in Motion," 223, **224**
Cooperative grouping guidelines, **134**
Creative thinking and learning, **31,** 89, 186
Critical thinking and learning, **31**
CTS (Curriculum Topic Study) guides, 209
Curiosity of children, xiii
Curricular approaches, 128–129

Index

Curriculum compacting, **134**
Curriculum Topic Study (CTS) guides, 209

D

Data analysis/probability, **147–148,** 228
Day-night cycle, 179, 181, **181**
Debrief the thinking process, **87**
Demonstrating student understanding, 7, **8, 11, 19, 26**. *See also* Assessments
    backwards design process for, 23–24, **26**
    for "Earth's Shape and Gravity," **176,** 180–185, **190**
    for "Matter and Its Transformations," **164, 167**
    for "The Flow of Matter and Energy in Ecosystems," 23–24, 30, 33, **34,** 73–74
    for "Understanding Changes in Motion," **210,** 213, **216–217**
Descriptive organizers, **116**
Designing Effective Science Instruction: What Works in Science Classrooms, **25**
Designs for Science Literacy, **25**
Developing Visual Literacy in Science, K–8, **54, 112**
Diagrams, **112**
Dialogue, 136
Differentiated Instructional Strategies for Science, Grades K–8, **54,** 154
Differentiated teaching, xvi, 128–141
    content differentiation, 130–131
        for "Earth's Shape and Gravity," 200
        for "The Flow of Matter and Energy in Ecosystems," **131**
        for "Understanding Changes in Motion," 223, **224**
    process differentiation, 131–136, **134–135**
        for "Earth's Shape and Gravity," 200, **201–202**
        grouping practices, 132–133
        learning activities, 132
        modification based on learner needs, 133–135
        resources, 132
        for "The Flow of Matter and Energy in Ecosystems," **136–138**
        time and space modifications, 133–135
        for "Understanding Changes in Motion," 225, **225–226**
    product differentiation, 136, 138, 139–140
        for "Earth's Shape and Gravity," 202
        for "Understanding Changes in Motion," 226
    re-teaching options, 128, 130, 141, **142–143**
    resources on, 154–155
Differentiating for the Young Child: Teaching Strategies Across the Content Areas, PreK–3, 155
Digital libraries, 80
Discourse, large- and small-group, **51, 56, 57, 103–105,** 192
Drawing Out Thinking (Instructional Tool 2.9), **44, 45, 121–124,** 191, 215, 217
Drawings, **44,** 48, **52,** 56, **121–122,** 191–192
Duschl, R. A., 4
Dynamic models, **43, 52,** 113

E

"Earth's Shape and Gravity," xiv, xvi, 4, 14, 171–204
    application of predictive phase to, 175–187
        identify conceptual target, 177
        identify criteria for determining student understanding, **176,** 180–185, **181–185**
        identify learning targets, **176,** 178–180, **180**
        inquiry and metacognitive goals and strategies, 186–187
        planning template, **175–176**
        unpack standards and identify concepts, knowledge, skills, and vocabulary, 177–178, **178**
    application of responsive phase to, 187–199
        determine formative assessments, 192
        lessons 1 to 7, 193–199, **197**
        planning template, **187–188**
        research children's misconceptions common to this topic, 172–173, 177–178, **178,** 189
        select sense-making strategies, **188, 190,** 191–192
        select strategies to elicit and confront your students' preconceptions, **188, 190,** 191
        select strategies to identify your students' preconceptions, **187,** 189, **190,** 191
        strategy selection template, **190**
    build your library on, 204
    considerations across the grades, 204
    overview of, 172–173
    reflection on, 199–200
    responding to needs of all learners, 200–202
        differentiating content, 200
        differentiating process, 200, **201–202**
        differentiating product, 202
    ties to literacy and numeracy, 202–203, **203–204**
    why this topic was chosen, 173–175
Ecosystems. *See* "The Flow of Matter and Energy in Ecosystems"
Education Oasis website, **54**
Elementary science kits, 206
Elementary Science Study (ESS), 206
Emotional learning system, xvi, 128, 156
ESS (Elementary Science Study), 206
Everyday Assessment in the Science Classroom, **26,** 175
Everyday Science Mysteries, **96**

F

Facet-based instruction, 215
Fairness routines, **84**
Five natural learning systems, xvi, 128, 156
5E Instructional Model, 10, **10–11**
Flat Earth Society, 197–198, 202
Flowcharts, **116**
Food chains, 5, **6,** 21
Force and motion. *See* "Understanding Changes in Motion"
Force and Motion: Stop Faking It!, 230
Formative assessments, 7, **19,** 24, **26, 42–44**
    determination of, 55, **55–56**
    differentiation of, 138
    for "Earth's Shape and Gravity," 192
    hints and resources for, **58**
    for "Matter and Its Transformations," 166, **167**
    probes, **91–92,** 217–218
    for "The Flow of Matter and Energy in Ecosystems," 24, 55–56, **55–56,** 58, 73–74
    for "Understanding Changes in Motion," 217–218

Index

FOSS (Full Option Science System), 128
Foundational knowledge of students, **5,** 6, 7, 14
Four-corner synectics, **126**
FreeMind, **115**
Full Option Science System (FOSS), 128

G

Gases. *See* "Matter and Its Transformations: Gas Is Matter"
Geometry, **146–147,** 202–203
Google Earth, **80**
Graphic models, **112,** 191–192
Graphic Organizer website, **54**
Graphic organizers, 31, **86.** *See also* Visual Tools
    resources on, **54, 117**
    task-specific, **44,** 53, **57, 116–117**
    work-plan organizer with fieldwork example, **32**
Graphing software, **112**
Gravity. *See* "Earth's Shape and Gravity"
Grouping practices, differentiation of, 132–133, **134, 137,** 200, **201, 225**
Guided inquiry, 27–28

H

Habits of mind, 143
Hand washing, 70
Hands-on experiments and activities and manipulatives, **44, 57, 125**
Hard-to-Teach Biology Concepts: A Framework to Deepen Student Understanding, xiii, 4, **124**
Hard-to-teach concepts, xiv, xvi, 4, 14
    "Earth's Shape and Gravity," 171–204
    "Matter and Its Transformations," 159–170
    "The Flow of Matter and Energy in Ecosystems," 17–126
    "Understanding Changes in Motion," 205–230
    why some topics are hard to teach, xv, **5,** 5–6, 207–208
Hints and resources, xiv, 7
    for planning in stage I to stage IV, **24–26**
    for planning in stage V, **33**
    for planning in stage VI, **36–37**
    for planning in stage VII, **50**
    for planning in stage VIII and stage IX, **54**
    for planning in stage X, **58**
Homework, targeted, 135, **135**
How Students Learn: Science in the Classroom, xiv

I

iMindMap, **115**
Independent study, **143**
Informational text strategies, **48, 99–101,** 165
Inhelder, B., 206
Inquire Within: Implementing Inquiry-Based Science Standards in Grades 3–8, **33**
Inquiry, xvi, 9, 15, 75
    in development of language literacy, 150
    differentiation and, 130–131
    goals for "Earth's Shape and Gravity," 186–187
    guided, 27–28
    hints and resources for planning in stage V, **33**
    open, 27–28, 186
    steps to identify focus of, 28–30
    strand map for "Scientific Inquiry: Evidence and Reasoning in Inquiry," **29**
    structured, 27–28
    Teaching the Five Essential Features of, 28, **76–82,** 164
    for "The Flow of Matter and Energy in Ecosystems," 26–30
"Inquiry and Scientific Explanations: Helping Students Use Evidence and Reasoning," **96**
Inquiry and the National Science Education Standards: A Guide for Teaching and Learning, **33**
Inspiration Software, **78, 80,** 99, **106, 115, 120**
Instructional approach, xvi, 18
Instructional Planning Framework, xiv, xv, 4, 6–9, **8,** 18
    applied to "Earth's Shape and Gravity," 171–204
    applied to "Matter and Its Transformations," 159–170
    applied to "The Flow of Matter and Energy in Ecosystems," 17–126
    applied to "Understanding Changes in Motion," 205–230
    compared with other models, 9–12, **10–11**
    differentiation of, xvi, 128–141
    implementation of, 12
        at elementary level, 127–156
        ten-stage process for, 18, **19**
        variations in third, fourth, and fifth grades, 153
    personal pre-assessment for use of, 12, **13**
    predictive phase of, 7, **8, 10,** 12, 18, **19**
    reflection on, 58, **59**
    research basis of, xiii, 4, 14
    responsive phase of, 7–9, **8, 10–11,** 12, 18–19, **19**
    ties to literacy and numeracy, 143–153, **145–149, 151–152**
Instructional strategies, xvi, 9, 12–13
    definition of, 18
    selection of, 18, **19**
        template for (*See* Strategy selection template)
Instructional Strategy Selection Tool (Instructional Tool 2.3), 41, **42–44,** 165, 189
    linguistic representations of knowledge, 41, **42–43,** 44
        language literacy and, 150
        Reading to Learn, **43,** 44, 53, **98–102,** 130, 150
        Speaking to Learn, **43,** 44, **103–106,** 150, 191
        Writing to Learn, **42,** 44, 51, **90–97,** 150
    nonlinguistic representations of knowledge, 41, **43–44,** 45, **45**
        Drawing Out Thinking, **44, 45, 121–124,** 191, 215, 217
        Kinesthetic Strategies, **44, 45, 125–126,** 215–217
        Six Kinds of Models, **43, 45, 107–113,** 191
        Task-Specific Graphic Organizers, **44,** 53, **57, 116–117**
        Thinking-Process Maps, **44, 51, 55, 57, 102, 118–120**
        Visual Tools, **44, 45, 114–120**
Instructional Tools, xiv, xv, xvi, 6, 12, 18, **50,** 132
    Drawing Out Thinking (2.9), **44, 45, 121–124,** 191, 215, 217
    Instructional Strategy Selection Tool (2.3), 41, **42–44,** 165, 189
    Kinesthetic Strategies (2.10), **44, 45, 125–126,** 215–217
    Reading to Learn (2.5), **43,** 44, 53, **98–102,** 130, 150
    Six Kinds of Models (2.7), **43, 45, 107–113,** 191
    Speaking to Learn (2.6), **43,** 44, **103–106,** 150, 191
    Teaching the Five Essential Features of Inquiry (2.1), 28, **76–82,** 164
    Three Strategies That Support Metacognition (2.2), 30, 31, **83–89,** 131, 164
    Visual Tools (2.8), **44, 45, 114–120**
    Writing to Learn (2.4), **42,** 44, 51, **90–97,** 150
Integrating Differentiated Instruction and Understanding by Design, 154, 156

Index

Integrating Science with Mathematics and Literacy: New Visions for Learning and Assessment, 155
Intelligences, 136, **138**
Interdisciplinary teaching, 128, 143
    resources for, 155–156
Interviews, 136

J

Just Read Now website, **99, 101, 102, 110**

K

Karplus, Robert, 206
Kinesthetic Strategies (Instructional Tool 2.10), **44, 45, 125–126,** 215–217
Koba, Susan, 4
KWL strategy, 165

L

Laboratory practicals, 136
Language literacy, relation of science to, 143, 150–153, **151–152**
    for "Earth's Shape and Gravity," 203, **203–204**
    for "Matter and Its Transformations," 170
    for "The Flow of Matter and Energy in Ecosystems," **145**
    for "Understanding Changes in Motion," 229–230
Large- and small-group discourse, **51, 56, 57, 103–105,** 192
Latex allergy, 166
Learner needs, modification based on, 133–135, **134, 138,** 200, **202, 226**
Learning
    cognitive, xvi, 30, 75, 128, 156
    conceptual, 75
    creative thinking and, **31**
    critical thinking and, **31**
    emotional, xvi, 128, 156
    inquiry-based, xvi, 9, 15
    metacognitive, xiv, xvi, **19,** 30, **31,** 75, 128
    physical, xvi, 156
    reflective (metacognitive), xvi, 128, 156
    research on, 74–75
    social, xvi, 128, 156
Learning, Creating, and Using Knowledge: Concept Maps as Facilitative Tools in Schools and Corporations, **120**
Learning activities, differentiation of, 132, **137,** 200, **201, 225**
Learning contracts, **134**
Learning goals and subgoals, 14, **19,** 20, **25.** *See also* Learning targets
Learning logs, 50, **51, 88, 90–91**
Learning progressions, 207

Learning sequence, 7, 14, 18, **19,** 20, **129,** 129–130
    for "Earth's Shape and Gravity," **180**
    for "Matter and Its Transformations," **163**
    for "The Flow of Matter and Energy in Ecosystems," 23, **23**
    for "Understanding Changes in Motion," 213
Learning styles, 129, 130, 132, 133
Learning targets, 7, **8,** 14, **19, 25**
    for "Earth's Shape and Gravity," **176,** 178–180, **180, 190**
    for "Matter and Its Transformations," 161, **163, 164, 167**
    for "The Flow of Matter and Energy in Ecosystems," 22–23, **22–23, 35, 46,** 48, **57**
    for "Understanding Changes in Motion," **210–211,** 213, **216, 218–219**
Lesson Plans, Inc., **125, 126**
Lesson products, differentiation of, 136, 138, **139–141,** 202
Life science lesson. *See* "The Flow of Matter and Energy in Ecosystems"
Linguistic representations of knowledge, 41, **42–43,** 44
    language literacy and, 150
    Reading to Learn, **43,** 44, 53, **98–102,** 130, 150
    Speaking to Learn, **43,** 44, **103–106,** 150, 191
    Writing to Learn, **42,** 44, 51, **90–97,** 150

M

Making Sense of Secondary Science: Research Into Children's Ideas, **25, 36, 37**
Math: Stop Faking It!, **107**
Mathematical literacy, 144
Mathematical models, **43, 107**
Mathematical power, 144
Mathematics, relation of science to, xv, xvi, 143–150
    common core standards for mathematics (grades 3–5) with connections to science, **149, 229**
    for "Earth's Shape and Gravity," 202–203
    mathematics standards supported by science instruction, 146–148
    for "Matter and Its Transformations," 170
    for "The Flow of Matter and Energy in Ecosystems," **145**
    for "Understanding Changes in Motion," 227, **227–229**
"Matter and Its Transformations: Gas Is Matter," xiv, xvi, 4, 14, 159–170
    application of predictive phase to, 161–164
        learning targets and sequence, **163, 164, 167**
        planning template, **163–164**
        science standards, 161, **162**
    application of responsive phase to, 164–169
        determine formative assessments, 166, **167**
        lessons 1 to 3, 166–169
        research children's misconceptions common to this topic, 165, **165**
        safety note for, 166
        select sense-making strategies, 166, **167**
        select strategies to elicit and confront your students' preconceptions, 166, **167**
        select strategies to identify your students' preconceptions, 165–166
        strategy selection template, 166, **167**
    build your library on, 170
    consideration across the grades, 170
    overview of, 159
    reflection on, 169
    ties to numeracy and literacy, 170
    why this topic was chosen, 159–161
MBLs (microcomputer-based laboratories), **107**
McDermott, Lillian, 206, 221
McTighe, J., 128
Measurement, **147, 227**
Mental models of Earth, 179–185, **181–185**
Metacognition, xiv, xvi, **19,** 75
    definition of, 30
    differentiation and, 130–131
    hints and resources for planning in stage V, **33**
    metacognitive approaches, 30, **31**
    metacognitive goals for "Earth's Shape and Gravity," 186
    reflection strategies for enhancement of, **101–102**
    for "The Flow of Matter and Energy in Ecosystems," 30–33
    Three Strategies That Support Metacognition (Instructional Tool 2.2), 30, 31, **83–89,** 131, 164
Metaphors, **43, 111**

Metaphors and Analogies: Power Tools for Teaching Any Subject, 110, 111
Microcomputer-based laboratories (MBLs), **107**
Mind mapping, **114–115**
Minstrell, J., 215
Misconceptions of students, 15–16, 19. *See also* Preconceptions of students
  about gas is matter, 165, **165**
  about motion, 207, 209, 215
  about spherical Earth, 172–173, 177–178, **178**, 189
  about the flow of matter and energy in ecosystems, 36, **38–40**
Model-It, **113**, **120**
Models
  Conceptual Change Model, 9, **10–11**
  dynamic, **43**, **52**, **113**
  for "Earth's Shape and Gravity," 178–179, 191–192
  5E Instructional Model, 10, **10–11**
  graphic, **112**, 191–192
  mathematical, **43**, **107**
  physical, **43**, **108**
  Six Kinds of Models, **43**, **45**, **107–113**, 191
  students' mental models of Earth, 179–185, **181–185**
  verbal
    analogies, **43**, **109–110**
    metaphors, **43**, **111**
  visual, **43**, **52**, **112**
Motion. *See* "Understanding Changes in Motion"
Multimedia projects, 136

N

Naive conceptions of students, 15, 16. *See also* Preconceptions of students
National Council of Teachers of Mathematics (NCTM), 144, 202
National Library of Virtual Manipulatives, **107**
National Research Council (NRC), 186, 208
National Science Digital Library (NSDL), 36, **36**, **115**, **120**
National Science Education Standards (NSES), xiv, 20, **25**, 209
  related to "Earth's Shape and Gravity," 175, 177–178
  related to "Matter and Its Transformations," 161, **162**
  related to "The Flow of Matter and Energy in Ecosystems," 20, **21**
  related to "Understanding Changes in Motion," 211–212

National Science Teachers Association (NSTA) resources, **24**
  Science and Children, 169
  science trade books for students, 101, 154
Natural curiosity of children, xiii
NCTM (National Council of Teachers of Mathematics), 144, 202
Negotiated assessment, 138, **139**
  feedback form for, **140**
Newton, Sir Isaac, 174, **197**, 207, 209
Newton's Ask a Scientist website, **81**
Next Generation Science Education Standards, 4, 20, 207
Nonlinguistic representations of knowledge, 41, **43–44**, 45, **45**
  Drawing Out Thinking, **44**, **45**, **121–124**, 191, 215, 217
  Kinesthetic Strategies, **44**, **45**, **125–126**, 215–217
  Six Kinds of Models, **43**, **45**, **107–113**, 191
  Task-Specific Graphic Organizers, **44**, 53, **57**, **116–117**
  Thinking-Process Maps, **44**, **51**, **55**, **57**, **102**, **118–120**
  Visual Tools, **44**, **45**, **114–120**
NoteStar tool, 87
Novak, Joseph, 172, 177, 180
NSDL (National Science Digital Library), 36, **36**, **115**, **120**
NSES. *See* National Science Education Standards
NSTA. *See* National Science Teachers Association resources
Numbers and operations, **146**, **227**
Nussbaum, Yossi, 172, 177, 180

O

Open-Ended Questioning: A Handbook for Educators, 91, **105**, **106**
Open inquiry, 27–28, 186
Options Diamond, **89**
Outdoor Science: A Practical Guide, 155

P

Paired problem solving, **85**
Peel the Fruit tool, **86–87**
Peer and self-assessment, **42–44**, 55
Personal agendas, 135, **135**, **143**
Photosynthesis, **21**
Physical learning system, xvi
Physical models, **43**, **108**
Physical movements and gestures, **44**, **56**, **57**, **126**
Physical science lessons. *See* "Matter and Its Transformations: Gas Is Matter"; "Understanding Changes in Motion"
Physics by Inquiry, 206
Physics by Inquiry, Volume 1, 221

Physics Education Group, 206, 221
Piaget, Jean, 206
Pictures, **112**
Plan and self-regulate, 31, **86–87**
Planning template: predictive phase, 33, **232**
  for "Earth's Shape and Gravity," **175–176**
  for "Matter and Its Transformations: Gas Is Matter," **163–164**
  for "The Flow of Matter and Energy in Ecosystems," 28, **34–35**
  for "Understanding Changes in Motion," 209, **210–211**
Planning template: responsive phase, **234–235**
  for "Earth's Shape and Gravity," **187–188**
  for "The Flow of Matter and Energy in Ecosystems," 59, **60–66**
  for "Understanding Changes in Motion," **218–219**
Portfolios, 136, 138, **140**
Posner, G. J., 9
Preconceptions of students, **5**, 6–7, 14–15. *See also* Misconceptions of students
  definition of, 15
  eliciting and confronting of, 8–9, **8**, **10**, 15, **19**, **42–44**
    for "Earth's Shape and Gravity," **188**, **190**, 191
    for "Matter and Its Transformations," 166, **167**
    selecting strategies for, 50–52, **51–52**, **57**
    for "The Flow of Matter and Energy in Ecosystems," **60–66**, 67–72, **69**, **71**
    for "Understanding Changes in Motion," 215, **216**
  identification of, 8, **8**, **10**, 15, **19**, **42–44**
    for "Earth's Shape and Gravity," **187**, 189, **190**, 191
    for "Matter and Its Transformations," 165–166, **167**
    selecting strategies for, 47, **48–49**, **57**
    for "The Flow of Matter and Energy in Ecosystems," 47, **49**, **60**, 67
    for "Understanding Changes in Motion," 215, **216**
  terminology for, 15–16
Predictive phase of Instructional Planning Framework, 7, **8**, **10**, 12, 18

Index

applied to "Earth's Shape and Gravity," 175–187
applied to "Matter and Its Transformations," 161–164, **163–164**
applied to "The Flow of Matter and Energy in Ecosystems," 20–33, **34–35**
applied to "Understanding Changes in Motion," 209–214
hints and resources for planning in stage I to stage IV, **24–26**
hints and resources for planning in stage V, **33**
planning template for, 33, **232**
role of classroom teachers in, 7, 18, 24
stages of, **19**
Prerequisites for units or lessons, 26
Principles and Standards for School Mathematics, 144
Probes, **91–92,** 217–218
Problem-solution organizers, **117**
Problem solving, mathematical, **148**
Process differentiation, 131–136, **134–135**
   for "Earth's Shape and Gravity," 200, **201–202**
   grouping practices, 132–133
   learning activities, 132
   modification based on learner needs, 133–135
   resources, 132
   for "The Flow of Matter and Energy in Ecosystems," 131–135, **134–138**
   time and space modifications, 133–135
   for "Understanding Changes in Motion," 225, **225–226**
Process organizers, **116**
Product, differentiation of, 136, 138, **139–140,** 202
Product differentiation, 136, 138, **139–140**
   for "Earth's Shape and Gravity," 202
   for "The Flow of Matter and Energy in Ecosystems," 138, **139–141**
   for "Understanding Changes in Motion," 226
Professional development, xiii
Professional learning community, 24
Project 2061, **36**
Proportional reasoning, 208

Q

Quality Questioning: Research-based Practice to Engage Every Learner, **105,** 106
Questioning, **105–106**
Questions, Claims, and Evidence: The Important Place of Argument in Children's Science Writing, **96,** 155

R

Ratio reasoning, 208–209
Re-teaching, 128, 130, 141, **142–143**
Readiness to learn, 133, **134**
Reading skills, 150–153, **151,** 229. *See also* Language literacy
Reading to Learn (Instructional Tool 2.5), **43,** 44, 53, **98–102,** 130, 150
Ready, Set, Science! Putting Research to Work in K–8 Science Classrooms, xiv–xv
Reciprocal teaching, **85–86**
Reflection strategies, **101–102**
Reflective learning system, xvi. *See also* Metacognition
Relational organizers, **117**
Representation, mathematical, **148**
Research on learning, 74–75
Resources for students, differentiation of, 132, **137,** 200, **201,** 225
Responsive phase of Instructional Planning Framework, 7–9, **8, 10–11,** 12, 18–19
   applied to "Earth's Shape and Gravity," 187–199
   applied to "Matter and Its Transformations," 164–169
   applied to "The Flow of Matter and Energy in Ecosystems," 35–58
   applied to "Understanding Changes in Motion," 220–223
   planning template for, **234–235**
   reflection on, 58, **59**
   stages of, **19**
Responsive teaching, 128, 129. *See also* Differentiated teaching
Rubrics, 86

S

Safety precautions
   latex allergy, 166
   live organisms in classroom, 70
Sagan, Carl, xiii
SAPA (Science: A Process Approach), 206
SCENE (Southwest Center for Education and the Natural Environment) website, **76, 106**
Schweingruber, H. A., 4
Science: A Process Approach (SAPA), 206
Science and Children, 169
Science Curriculum and Improvement Study (SCIS), 206
Science Curriculum Topic Study: Bridging the Gap Between Standards and Practice, **24**
Science Education Resource Center's Using Data in the Classroom, **80**
Science Formative Assessment: 75 Practical Strategies for Linking Assessment, Instruction, and Learning, **58, 122, 124**
Science literacy, 14
Science Matters: Achieving Science Literacy, **24**
Science notebooks, 53, **55, 57,** 90, **93–94,** 132, 136, 152
Science Notebooks: Writing About Inquiry, **94,** 156
Science standards, xiv, 4, 14, 18, 20, **25**
   related to "Earth's Shape and Gravity," 175, 177–178
   related to "Matter and Its Transformations," 161, **162**
   related to "The Flow of Matter and Energy in Ecosystems," 20, **21**
   related to "Understanding Changes in Motion," 211–213
Science teachers, xiv
   content knowledge of, 7
   personal pre-assessment of, 12, **13**
   professional development of, xiii
   resources for (*See* Build your library; Hints and resources)
   role in predictive phase of Instructional Planning Framework, 7, 18, 24
Science Technology Concepts (STC), 128
Science ties to literacy and numeracy, 143–153, **145–149, 151–152**
   for "Earth's Shape and Gravity," 202–203, **203–204**
   for "Matter and Its Transformations," 170
   for "The Flow of Matter and Energy in Ecosystems," **145**
   for "Understanding Changes in Motion," **227–229,** 227–230
Science trade books for students, 101, 154, 170, 204, 230
Science Writing Heuristic (SWH), **51,** 53, **96–97**
Scientific method, xiv
SciLinks
   food chains (K–4), 154
   food webs (K–4), 154
   foods and energy (5–8), 154
   force and motion (K–4), 230
   force and motion: Newton's first law (5–8), 230

Index

force and motion: position and motion (5–8), 230
force of gravity (K–4), 199
forces (K–4), 230
gravity (5–8), 199
gravity and orbiting objects (K–4), 199
how do organisms get energy? (5–8), 154
matter and energy (5–8), 154
matter and energy (K–4), 154
matter and gravity (5–8), 199
physical and chemical properties of matter (5–8), 169
properties of matter (K–4), 169
states of matter (5–8), 169
states of matter (K–4), 169
SCIS (Science Curriculum and Improvement Study), 206
SCORE website, **54**
Seasonal Partners, **134**
Seeds of Science: Roots of Reading, **99**, **101**, **105**, 155
Self-evaluation, **88**, 131
Self-regulated thinking, 30, 31, **31**, **84–85**, 131, 217
Sense-making strategies, **8**, 9, **11**
 connecting information, **11**
 for "Earth's Shape and Gravity," **188**, **190**, 191–192
 for "Matter and Its Transformations," 166, **167**
 perceiving, interpreting, and organizing information, **11**
 retrieving, extending, and applying information, **11**
 selection of, **19**, **42–44**, 52–53, **57**
 for "The Flow of Matter and Energy in Ecosystems," **60–64**, **66**, 72–73, **74**
 for "Understanding Changes in Motion," 215, **216**, 217
Sequential organizers, **116**
Shayer, Michael, 208
Shouse, A. W., 4
Six Kinds of Models (Instructional Tool 2.7), **43**, **45**, **107–113**, 191
Smart Boards, **115**, **117**, **120**
Social interaction, 12
Social learning system, xvi, 128, 156
Southwest Center for Education and the Natural Environment (SCENE) website, **76**, **106**
Space modifications, 133, 135, **135**, **138**, 200, **202**, **226**
Speaking skills, 150–153, **152**, 229–230. *See also* Language literacy
Speaking to Learn (Instructional Tool 2.6), **43**, 44, **103–106**, 150, 191
Spectroscopy, 179
Speed Comparison Task, 207, 221–222, **222**, **224**

Spherical Earth and gravity concept. *See* "Earth's Shape and Gravity"
Stations, **142**
STC (Science Technology Concepts), 128
Stepping Up to Science and Math: Exploring the Natural Connections, **112**
Stop Faking It! series, **24**, **107**, 230
Strand maps
 in *Atlas of Science Literacy*, **25**, 28
 "Scientific Inquiry: Evidence and Reasoning in Inquiry," 29
 National Science Digital Library resources for, **36**
Strategies That Work: Teaching Comprehension for Understanding and Engagement, **101**
Strategy selection template, 56, **233**
 for "Earth's Shape and Gravity," **190**
 for "Matter and Its Transformations," 166, **167**
 for "The Flow of Matter and Energy in Ecosystems," **46**, **57**
 for "Understanding Changes in Motion," **216–217**
Strike, K. A., 9
Structured inquiry, 27–28
Student questioning, **105–106**
Student VOC (Vocabulary) Strategy, **98–99**
Summary frames, 95
Summative assessments, 7, 24, **26**, 138
SWH (Science Writing Heuristic), **51**, 53, **96–97**
Symbolic tools, **5**, 6
Systems diagrams, **118–120**

T

Taking Science to School: Learning and Teaching in Grades K–8, xv, 208
Talk about thinking, 31, **85–86**
Targeted homework, 135, **135**
Task-specific graphic organizers, **44**, 53, **57**, **116–117**
TeacherVision, **110**
Teaching, Learning & Assessing Science 5–12, **91**
Teaching Constructivist Science: Nurturing Natural Investigators in the Standards-Based Classroom, 155
Teaching Reading in Science, **99**, **101**, **102**, 155
Teaching Science Through Inquiry With Archived Data, **78**
Teaching the Five Essential Features of Inquiry (Instructional Tool 2.1), 28, **76–82**, 164
Teaching-with-Analogies Model, **110**
Ten-stage process for implementing Instructional Planning Framework, 18, **19**
THC strategy, 165–166
The Comprehension TookKit: Language and Lessons for Active Literacy, **101**
The Everyday Science Mysteries: Stories for Inquiry-Based Science Teaching, 155
"The Flow of Matter and Energy in Ecosystems," xiv, xvi, 4, 14, 17–126
 application of predictive phase to, 20–33
 hints and resources for, **24–26**, **33**
 identify and sequence subgoals, 23, **23**
 identify conceptual target, 20, **21**
 identify criteria for determining student understanding, 23–24
 identify inquiry and metacognitive goals and strategies, 26–33, **27**, **29**, **31–33**
 planning template, 33, **34–35**
 unpack standards and identify concepts, knowledge, skills, and vocabulary for content, 22, **22**
 application of responsive phase to, 35–58
 activity selection and sequencing, 35
 determine formative assessments, 24, 55–56, **55–56**, **58**, 73–74
 hints and resources for, **36–37**, **50**, **54**, **58**
 Instructional Strategy Selection Tool, 41, **42–44**
 linguistic representations of knowledge, 41, **42–43**, 44, **90–106**
 nonlinguistic representations of knowledge, 41, **43–44**, **45**, **45**, **107–126**
 planning template, 59, **60–66**
 research children's misconceptions common to this topic, 36, **38–40**
 select sense-making strategies, 52–53, **54**, 72–73, **74**
 select strategies to elicit and confront your students' preconceptions, 50–52, **51–52**, **60–66**, 67–72, **69**, **71**

Index

select strategies to identify your students' preconceptions, 47, **48–50, 60,** 67
strategy selection template, 46, **46,** 57
choice board for, **141**
reflection on, 58, **59**
responding to needs of all learners, 130–138
differentiating content, 130–131, **131**
differentiating process, 131–135, **134–138**
differentiating product, 138, **139–141**
ties to literacy and numeracy, **145**
variations in third, fourth, and fifth grades, 153
The Parallel Curriculum: A Design to Develop High Potential and Challenge High-Ability Learners, 154
The Teacher's Guide to Leading Student-Centered Discussions: Talking About Texts in the Classroom, **105**
Thinking maps, **118–120**
Thinking Maps website, **115**
Thinking-process maps, **44, 51, 55, 57, 102, 118–120**
ThinkingMaps software, **120**
Three Strategies That Support Metacognition (Instructional Tool 2.2), 30, 31, **83–89,** 131, 164
Tiered activities, **135**
Tiered product assignments and tasks, 138, **139**
Time modifications, 133–135, **135, 138,** 200, **202, 226**
Tips for the Science Teacher: Research-based Strategies to Help Students Learn, 54
Tomlinson, C. A., 128
Transmissive writing, **94**
Trowbridge, David, 221
Truth routines, **83**
Tweed, Anne, 4

U

Uncovering Student Ideas in Physical Science, Vol. 1: 45 New Force and Motion Assessment Probes, 218, 222, 230
Uncovering Student Ideas in Science, **37, 58, 90, 91, 92, 96,** 156
Understanding by Design, 23–24, **26,** 175, 209
"Understanding Changes in Motion," xiv, xvi, 4, 14, 205–230
application of predictive phase to, 209–214
determine formative assessments, 217–218
identify and sequence learning targets, **210–211,** 213, **216**
identify conceptual target, 211
identify criteria for determining student understanding, **210,** 213, **216–217**
planning template, 209, **210–211**
research children's misconceptions common to this topic, 215
select sense-making strategies, 215, **216,** 217
select strategies to elicit and confront your students' preconceptions, 215, **216**
select strategies to identify your students' preconceptions, 215, **216**
unpack standards and identify concepts, knowledge, skills, and vocabulary, 211–213
application of responsive phase to, 220–223
lessons 1 to 4, **220,** 220–223, **222**
planning template, **218–219**
build your library on, 230
consideration across the grades, 207, 230
Curriculum Topic Study guides related to, 209
overview of, 206–207
reflection on, 223
responding to needs of all learners, 223–226
differentiating content, 223, **224**
differentiating process, 225, **225–226**
differentiating product, 226
what makes these ideas difficult, 208–209
why this topic was chosen, 207–208
Understanding Models in Earth and Space Science, **108, 111**
Using Analogies in Middle and Secondary Science Classrooms, **110**

V

Venn-Euler diagrams, **116, 117**
Verbal models
analogies, **43, 109–110**
metaphors, **43, 111**
Verbal presentations, 136
Visible thinking, **83–84**
creativity routines, **89**
fairness routines, **84**
truth routines, **83**
Visible Thinking website, **33, 76, 79, 83–84, 89**
Visual models, **43, 52, 112**
Visual Tools (Instructional Tool 2.8), **44, 45, 114–120**
Vocabulary development, **25, 98–99,** 130
for "Earth's Shape and Gravity," 179
for "The Flow of Matter and Energy in Ecosystems," 22, **22, 34**
VoiceThread, **86, 88, 104**
Vosniadou, Stella, 179, 180

W

Work-plan organizer
with fieldwork example, **32**
for "The Flow of Matter and Energy in Ecosystems," **71**
Writing in Science: How to Scaffold Instruction to Support Learning, 156
Writing scientific explanations, **94–96**
Writing skills, 150–153, **151,** 229. *See also* Language literacy
Writing to Learn (Instructional Tool 2.4), **42,** 44, 51, **90–97,** 150